# WEAPONS

# A PICTORIAL HISTORY

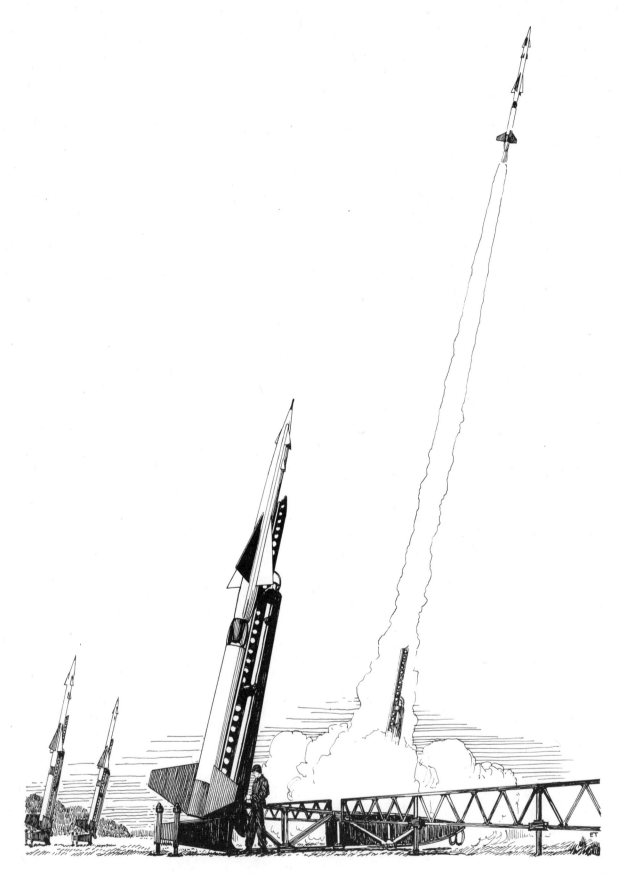

THE ARMY'S "Nike," ANTIAIRCRAFT GUIDED MISSILE

# WEAPONS

## A PICTORIAL HISTORY

WRITTEN AND ILLUSTRATED BY

*Edwin Tunis*

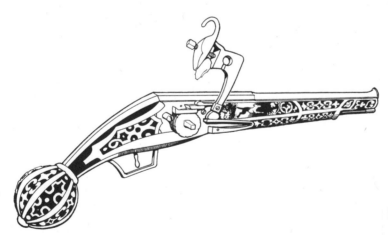

*BALTIMORE AND LONDON*

THE JOHNS HOPKINS UNIVERSITY PRESS

*To three old friends:*

GEDDY, BETTY AND BILLY HARDY

Hardcover edition published by The World Publishing Company,
Cleveland and New York. Copyright 1954 by Edwin Tunis
All rights reserved
Design and typography by Jos. Trautwein

Johns Hopkins Paperbacks edition, 1999
Printed in the United States of America on acid-free paper
3 5 7 9 8 6 4 2

The Johns Hopkins University Press
2715 North Charles Street
Baltimore, Maryland 21218-4363
www.press.jhu.edu

**Library of Congress Cataloging-in-Publication Data**

Tunis, Edwin, 1897–1973.
Weapons : a pictorial history / written and illustrated by Edwin Tunis.
p.    cm.
Originally published: Cleveland : World Pub. Co., 1954.
ISBN 0-8018-6229-9 (pbk. : alk. paper)
1. Weapons—History. 2. Armor—History. I. Title.
U800.T8   1999
623.4′09—dc21                                              99-20399

A catalog record for this book is available from the British Library.

# PREFACE

IT WOULD BE comforting to believe that *nobody* could have written this book "right out of his head"; certainly I couldn't do it. I hollered for help at every step and got it. It came from experts and from just plain folks-who-knew-something. They have loaned me rare books; they have loaned me valuable weapons; with admirable patience they have explained and demonstrated. If I have still failed to understand, that isn't their fault and they cannot be held accountable for the book's errors and shortcomings.

Brigadier General Frank D. Bowman, U.S.A.; Captain Elmer C. Clusman, U.S.N.R.; Mr. Osborn M. Curtis, Jr.; Mr. James V. Lecocq, of the Office of Technical Information, Army Ordnance; Mr. Harold I. Lessem, Acting Superintendent of Fort McHenry; Mr. F. D. McHugh, of the Office of Technical Information, Army Ordnance; Mr. Richard Harding Randall, Jr., of the Metropolitan Museum; and Mr. Eustis Walcott have given indispensable aid and comfort and I thank them gratefully.

Above and beyond all others I thank my wife Lib, who, with only the faintest interest in the subject, has typed all these words, and not complained too much, even when correction was itself corrected.

E. T.

*Long Last*

*February* 2, 1954

# CONTENTS AND ILLUSTRATIONS

"Nike," Antiaircraft Guided Missile  2

Preface  5

Note  13

*Stone Weapons*  15

Prehistoric Man Throwing a Tied Stone  15

Simple Sling  15

War Spear and Fishing Spear  16

Flint Knife  16

Polished Stone Ax  16

Chipped Flint Ax  16

Throwing Stick  17

American Indian Bow and Arrow  17

American Indian with Bow and Arrow  17

Types of Arrowheads  18

Filipino Blowgun  18

Australian Boomerang  18

*Copper and Bronze* (5000 B.C.–1000 B.C.)  18

Copper Ax  18

Chaldean Warriors 2000 B.C.  19

Bronze Chaldean Dagger  19

Copper Spearhead  19

Asiatic Composite Bow  19

Egyptian Bow and Arrow  20

Egyptian Sword  20

Chaldean War Chariot  21

Egyptian War Chariot  21

Babylonian Battering Spear  22

*The Greeks and Bronze* (1000 B.C.)  22

Homeric Warrior with Tower Shield  23

Homeric Warrior with Figure-eight Shield  23

Bronze Spearhead  23

Long Stabbing Sword  23

Homeric Foot Soldier  24

The Trojan Horse  24

*The Greeks and Iron* (680 B.C.–150 B.C.)  25

Greek Hoplite 500 B.C.  25

Peltast  26

Greek Phalanx  26

Diagram of Phalanx Advance  26

Corinthian Helmet  27

Boetian Helmet  27

Greek Sword  27

*The Roman Soldier* (200 B.C.–A.D. 400)  28

Legionary Helmet  29

Roman Centurion and Legionary  29

Sword  29

Velite  30

Gladiator's Helmet  30

*Roman Sieges and Siege Engines*
(200 B.C.–A.D 400)  30

Battering Ram, Testudo and Tower  31

Onager in Action  32

Onager Loaded, Showing Slip-Hook  33

Ballista  33

Catapult for Javelins  34

Trigger Operation of Catapult  34

*The Dark Ages*  35

Frankish Warrior, Sixteenth Century  35

Frankish Throwing Ax  35

Early Frankish Shield  35

Anglo-Saxon Taper Ax  36

Anglo-Saxon Broadax  36

Frankish Soldier, Ninth Century  36

Morning Star  37

Frankish Sword  37

Anglo-Saxon Iron Arrowheads  37

Scramasax or Dirk  37

| | |
|---|---|
| Anglo-Saxon Spearhead | 37 |
| Anglo-Saxon Bill | 37 |

### The Norman Conquest (1066)  38

| | |
|---|---|
| Two Saxons and Norman Knight | 38 |
| Norman Sword and Scabbard | 39 |
| Mace | 39 |
| Knights' Spurs | 40 |
| Chain-Mail Pattern | 40 |

### Castles  41

| | |
|---|---|
| Early Norman Castle | 41 |
| Stone Keep, Interior View | 42 |
| Tower Keep | 43 |
| Crossbowman | 43 |
| Archer Shooting through Crenel | 43 |
| Hoardings | 44 |
| Machicolations | 44 |
| Moated Castle, c. 1300 | 45 |
| Portcullis, Closed | 45 |

### War Games (1200–1300)  46

| | |
|---|---|
| Knights Jousting | 46 |
| Tilting at the Quintain | 47 |
| Bout with Quarterstaves | 48 |
| The "Exercise of the Sword-and-Buckler" | 48 |

### Knights and Armor (1200–1300)  48

| | |
|---|---|
| Iron *Heaume* | 49 |
| Helmets | 49 |
| Falchion and Misericord | 50 |
| Knight Arming for Tournament | 51 |
| Caltrops | 51 |

### Medieval Arms and "Gyns" (1300–1400)  52

| | |
|---|---|
| Foot Soldier | 52 |
| The Royal Standard | 53 |
| Crested *Heaume* | 53 |
| The "Mouse" | 54 |
| Scaling Ladder | 54 |
| Springal | 54 |
| Ballista | 55 |
| Mangonel or "Nag" | 55 |
| Large Trebuchet | 56 |

### Longbows and Crossbows (1300–1400)  57

| | |
|---|---|
| Archers, Fourteenth Century | 58 |
| Two-Fingered Draw | 59 |
| Longbow | 59 |
| Flight Arrow and Livery Arrow | 59 |
| Diagram: Aiming Arrows | 60 |
| Arrowheads | 60 |
| Shooting Tab and Bracer | 61 |
| Mantlet | 62 |
| Pavise | 62 |
| Lock Mechanism of Crossbow | 63 |
| Simple Crossbow | 64 |
| Archer Drawing Crossbow | 64 |
| Belt Claw for Drawing Crossbow | 64 |
| Quarrel for Crossbow | 64 |
| Operating Cord and Pulley of Crossbow | 64 |
| Composite Crossbow | 65 |
| Staff Sling | 65 |

### Knights and Guns (1300–1400)  66

| | |
|---|---|
| Basinet with Movable Visor | 66 |
| The Black Prince in Armor | 66 |
| Armored or *Barded* Horse | 67 |
| *Pot de Fer*—First Metal Cannon | 67 |
| Lance with *Coronel* Point | 67 |
| Cannon at Battle of Crécy | 68 |
| Early Bombard | 68 |
| Hooped Bombard | 69 |
| One-man Hand "Gonne" | 69 |
| Two-man Hand "Gonne" | 69 |
| Hand "Gonne," Fourteenth Century | 69 |

### Proof Armor, Arbalests and Breechloaders (1400–1500)  70

| | |
|---|---|
| Joust with Barrier | 70 |
| Helm, Fifteenth Century | 71 |
| Billman wearing Brigandine Jacket | 71 |
| Pole Arms: Oxtongue, Poleax, Glave | 72 |
| Hunter Shooting an Arbalest | 72 |
| Setting Arbalest with Windlass and Tackle | 73 |
| Gaffle for Setting Arbalest | 73 |

"Works" of a Gaffle 73
"Goat's Foot" 74
Prodd or Stonebow 74
Setting a Prodd 75
War Quarrels and Game Bolts 75
Two-man Culverin 75
One-man Culverin 75
Large Siege Bombard 76
Breechloader 77

Matchlocks and Wheel Locks (1500–1600) 78
Arquebusier and Helper 78
Matchlock Gun 79
Inside of Matchlock 79
Bullet Pouch and Touch-box 80
Musketeer Using Rest 80
Wooden Powder Chargers on Bandolier 80
The "Monk's Gun" 81
Inside of Wheel Lock 81
Spanner for Winding Wheel Lock 81
Ball-butted Wheel-lock Pistol 82
"Holy-water Sprinkle" 82

Soldiers (1500–1600) 83
Knight in Maximilian Armor 83
Reiter 84
Pikeman 84
Swiss Halberdier 85
Pole Arms: Guisarme, Partisan, Fauchard,
   Halberd 85
Flamberge 85
Claymore 86
Cinquedea 86
Duel with Swords and Daggers 86
Duel in the French Style 86
Rapier Hilts 87
Nobleman with Rapier in Baldric 87
Sixteenth-Century Headpieces 87

Cannons (1500–1600) 88
Ship Cannon, Sixteenth Century 88
The Six Cannon Sizes of Henry II 89

Moving a Culverin 90
"Double firing" from a Mortar 90

Cavaliers and Snaphances (1600–1700) 91
Dutch Pikeman of 1607 91
Cavalier and Attendant 92
Puritan Roundhead 92
Inside a Snaphance Lock 93
Inside a Flintlock 93
Plug Bayonet 94
Soldier with Flintlock Musket 94
Brass-barreled Blunderbuss 95
Coachman with Blunderbuss 95
Spring-gun 95

Field Guns and Bastioned Forts (1600–1700) 96
Swedish Cast-iron Four-pounder 96
Canister 96
Small Coehorn Mortar 97
Howitzer 97
Gunner's Quadrant and Level 97
Two Bastions of a Vauban Fort 98
Profile of a Vauban Fort 99
Diagram of Vauban's Siege System 99
Powder Magazine 100

The Kentucky Flintlock Rifle (1727–1820) 101
German Rifle 101
Frontiersman with Rifle 101
Kentucky Rifle, Left Side 102
Kentucky Rifle, Right Side 102
Loading Patched Bullet 102
Rifling Bench 103
Pouch and Powder Horn 104
Bullet Mold 104
The Hall Rifle 105
Court or Dress Sword 105
Naval Cutlass 105
Cavalry Saber and Scabbard 105
Flintlock Pistol 106

Eighteenth-Century Artillery (1700–1800) 106
Horse Artillery 106

| | |
|---|---|
| Grape Shot | 107 |
| Barbette Carriage | 107 |
| Naval Truck Gun | 108 |
| Ladle | 108 |
| Rammer | 109 |
| Handspike | 109 |
| Wormer | 109 |
| Sponge | 109 |
| Linstock | 109 |
| "Cat" | 109 |
| Scraper | 109 |
| Semi-fixed Ammunition | 110 |

### Percussion (1800–1850) · 110

| | |
|---|---|
| Forsyth "Scent-bottle" Percussion Lock | 110 |
| Inside of the Scent-bottle | 111 |
| Shaw Cap Lock | 111 |
| U.S. Army Model of 1841 Rifle | 112 |
| Belted Ball and Bore | 112 |
| Diagrams of Minié Ball | 112 |
| Maynard Tape Primer | 113 |
| Derringer | 113 |
| Pepperbox | 113 |
| "Texas Model" Colt's Patent Revolver | 114 |

### The Rockets' Red Glare and the "Soda Bottle" (1800–1850) · 114

| | |
|---|---|
| Eighteenth-century Bomb with Rings | 114 |
| Handling a Bomb with Tongs | 115 |
| English Military Rocket about 1900 | 115 |
| Friction Primer | 115 |
| Congreve Military Rocket | 115 |
| Foredeck of a Bomb Ketch | 116 |
| Cavelli Shot | 116 |
| Whitworth Shot | 116 |
| Dahlgren "Soda Bottle" | 117 |

### Gastight Cartridges and Smokeless Powder (1850–1900) · 117

| | |
|---|---|
| Maynard Carbine and Pierced Cartridge | 117 |
| Pistol-carbine | 118 |
| Model One, Smith and Wesson Revolver | 118 |
| Navy Single-shot Pistol | 118 |
| Krag-Jorgenson Rifle | 119 |
| Henry Repeating Rifle | 119 |
| .30 Caliber Cartridge for Krag Rifle | 119 |
| Borchardt Automatic Pistol | 120 |

### Rifled Cannon and Recoil Mechanisms (1850–1900) · 120

| | |
|---|---|
| Light Parrott Rifled Cannon | 120 |
| Shell for Parrott Rifle | 121 |
| Thirteen-inch Civil War Mortar | 121 |
| Breechblock for Big Gun | 122 |
| Fixed Ammunition | 123 |
| Gun on Disappearing Carriage | 123 |
| French "75" in Recoil after Firing | 124 |

### Quick-Firing and Machine Guns (1850–1900) · 124

| | |
|---|---|
| Montigny Mitrailleuse | 125 |
| Early Gatling Gun | 125 |
| Gardner Portable Quick-firing Gun | 126 |
| Maxim Automatic Machine Gun | 126 |
| Colt-Browning Machine Gun | 126 |

### Shoulder Arms and Hand Arms (1900–1925) · 127

| | |
|---|---|
| U.S. Air Force Survival Gun | 127 |
| Double-barreled Eight-Gauge Elephant Gun | 127 |
| First Colt Automatic Pistol | 128 |
| M1903 Springfield Rifle | 129 |
| Lewis Light Machine Gun | 129 |

### Great Guns and Little Guns (1900–1925) · 130

| | |
|---|---|
| 37 mm. Gun on Tripod Mount | 130 |
| Truck-mounted Antiaircraft Gun 1918 | 130 |
| "Skysweeper" 75 mm. Radar-Controlled Antiaircraft Gun | 131 |
| 155 mm. Howitzer Pulled by Five-ton Artillery Tractor | 132 |
| Fourteen-inch Railway Gun | 133 |
| The Paris Gun | 133 |
| Armor-piercing Shell | 134 |
| Canister, 1953 | 135 |

| | | |
|---|---|---|
| *Special Weapons* (1900–1925) | 135 | |
| Da Vinci's "Tank" | 135 | |
| British Heavy Tank, World War I | 135 | |
| Pineapple Hand Grenade | 136 | |
| Rifle Grenade and Launcher | 136 | |
| 60 mm. Trench Mortar, Current Model | 137 | |
| Diagram of Torpedo | 138 | |
| Mine Detector | 139 | |
| Navy Frogman Clearing a Mine Field | 139 | |
| World War I Gas Mask | 140 | |
| *Self-Loading and Automatic Guns After* 1925 | 140 | |
| Garand Rifle | 140 | |
| "T-47" Automatic Rifle | 140 | |
| M1 Carbine | 141 | |
| Carbine with "Sniperscope" | 142 | |
| Tommy Gun | 142 | |
| M3 Submachine Gun | 143 | |
| Browning Air-cooled Machine Gun | **143** | |
| Sizes of .30- and .50-Caliber Rounds | 144 | |
| 1954 | 144 | |
| 155 mm. Gun on Pneumatic-tired Carriage | 144 | |
| 155 mm. Gun on Self-propelled Carriage | 145 | |
| "General Pershing" 45-ton Tank | 145 | |
| "Patton" 48-ton Medium Tank | 146 | |
| Walker Bulldog 26-ton Light Tank | 146 | |
| 280 mm. Mobile Atomic Gun | 147 | |
| 3.5-inch Super-bazooka | 147 | |
| 57 mm. Recoilless Rifle | 148 | |
| 75 mm. Recoilless Rifle | 148 | |
| V-1 Buzz Bomb | 149 | |
| V-2 Rocket Bomb | 149 | |
| Flame Thrower | 150 | |
| 2000-pound Demolition Bomb | 151 | |

ASIDE FROM those for hunting, there are two kinds of weapons as there are two kinds of fighting, offensive and defensive. An offensive weapon, whether it be a simple club or an automatic rifle, is an *arm* which will lengthen the reach of a man—something with which he hopes to strike from a greater distance than that from which he can be struck. The other kind of weapon, the defensive one, the man hopes will ward off blows which are directed at him; it may be a wooden shield, a suit of armor or a fort.

This book is mainly about offensive weapons but because one kind has usually been opposed to the other, the defensive ones are here too. The book is not about wars but about weapons. It has almost nothing to say about strategy and mutters only enough about tactics and siegecraft to explain the use of certain arms and "engines."

To avoid repetition, a weapon is discussed in detail only when it was of prime importance in its time, or when there is something new to say about it. For instance: The bow was first used before the memory of man; it made its most recent appearance as a "civilized" weapon of war in 1630; to examine its use in all the years between, when it was still a bow and still operating on the same old principle, would build a mountain of dullness. So it gets the full treatment only at its peak in history when its force tipped the balance and the longbow was the most important weapon in the world.

Another thing: men love weapons and collect them. Nearly every division of this book is the area of someone's fervent enthusiasm. "Someone" is going to feel that his pet subject is slighted. That's right, it is. This is a swift look at the whole business from start to finish. To cover in detail every known weapon, including the complicated ones of the present, would take not a book but a library. Merely to mention all the varieties would fatten this volume to dictionary size and there wouldn't be any room for pictures.

It seems customary for the author of a book on this kind of subject to announce at the start his deep-rooted abhorrence of all war; so this author does so, here and now. It would be hard to find any decent man who doesn't sincerely hate war and there are a lot of decent men in the world; yet mankind, controlled for the most part by these same decent men, has always had war and still has it. Further, if we are honest, nearly all of us who have hair on our chins or the prospect of it must admit to a built-in interest in fighting and the implements of fighting. So there we are, stuck with our human inconsistency: We hate war at the same time that it fascinates us!

Every device we have contrived in the past to shorten the lives of our fellowmen has been robbed of much of its terror and finally nullified by some countermeasure. There seems no reason to believe that this neutralizing process has stopped. Let's look at the record.

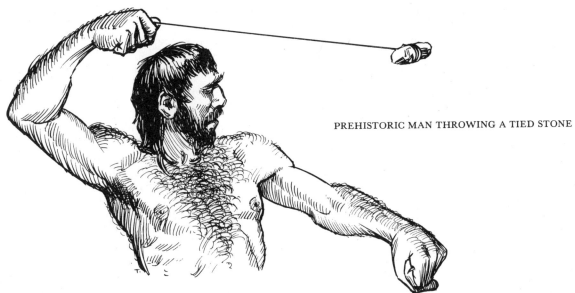

PREHISTORIC MAN THROWING A TIED STONE

## Stone Weapons

It's useless to talk about the kind of weapon which was seized merely by chance. Danger threatened, the hair on the spine of the ape man bristled and he grabbed the nearest stick and struck out with it. Baboons do as much.

What weapon did man first *make* for himself? That's something to guess at. Perhaps he found a rock which hefted well in his hand and kept it by him. When he tied a thong of hide or vine to that stone he made his first weapon. Maybe his original idea was only to keep the stone from being lost; but whirled at the end of its lanyard and let fly, it would outdistance anything his arm alone could do.

What he had was the crude beginning of a sling and all he needed to complete it was a way to hold on to the thong and let the stone go. That was easy: he made a loop on the end of the thong to slip over a finger, then doubled the thong and held the free end. The stone was slung in the bight. A cradle for the stone was an extra refinement.

The spear must have been next, though it may have been made ahead of the sling. At first it was only a stick sharpened by charring one end and rubbing it on a boulder. Soon it was given a head made with a naturally pointed stone lashed in a cleft in the stick; and shortly it was headed with a stone shaped, not by luck, but by the warrior himself, or by some member of his tribe who was good at shaping and could be paid for the job with a wolf skin. Flint proved to be the best material for heads, though other stones, and sometimes bone, were used.

Flint is not actually so difficult to shape as it once was thought to be. An expert (there are few) can make a complete, usable arrowhead in five minutes. Flint has the characteristic possessed by many hard, pure minerals, including plate glass, of chipping rather than splitting when it is struck an educated, glancing blow. It is even possible to chip these materials by hand pressure with a blunt-pointed bone implement, or a tooth, which is the way the finest flint articles were made. This can't be done casually by anybody who happens to pick up a rock and old bone, but for skilled hands it is by no means the dreary labor it was formerly thought to be.

Mr. H. L. Skavlem of Wisconsin studied the matter thoroughly and became so skilled that, using only such tools as the Indians had, he made arrow points as good as those any aborigine ever produced. In fact, he once demonstrated arrowhead making to a group of amazed Indians—who had never heard of such a thing!

Incidentally, the same gentleman also exploded the old theory that a stone ax required a lifetime to make. He produced one complete with edge and hafting groove and ground it to a smooth finish in four hours. Then he put a handle on it and used it to cut down a small tree.

Stone weapons didn't disappear when metal ones were invented. The two kinds existed together for centuries, as is proved by the finding of stone axes shaped to imitate bronze ones. Stone arrowheads and lance heads were used by the

SIMPLE SLING

Normans in France as late as the eighth century A.D.

From spearheads the flint shapers soon progressed to other weapons and implements such as axes and knives. Some axes were planned to be used without handles but most of them were hafted. All primitive people were skilled at lashing and knot tying. They could lash an ax head into a split handle and make it completely firm and dependable, but there was a better way to haft an ax. The flint head was forced into a split made in a growing tree limb and simply left there for two or three years. The tree, seeking to heal its wound, would fill in the split tightly around the stone and the finished ax could be harvested by simply cutting it off at both ends. Time was no object. A few late stone ax heads, or *celts* as they are called, have been found which have holes through them for the haft to be inserted. Some stone axes such as the American Indian tomahawk were intended for throwing.

in Swiss lake dwellings and there are drawings on cave walls which show them in use. They were much longer than the American Indian bows of more recent times and not very powerful.

The materials available determined what bows and arrows were made of. The American bows were made of Osage orange, hickory and ash; the strings were rawhide or animal tendons. Arrows were naturally straight, or artificially straightened; fairly strong and as light as it was practical to make them. Indians in eastern North America used viburnum, which is still called arrowwood. The heads of primitive arrows were usually of flint or bone. The nock, or string end of the shaft, had two or three half-feathers attached to it as an aid to accurate flight. Many primitive arrows had a lump at the nock to make the arrow easier to grasp between the thumb and finger. This is not the best way to draw a bow but it is the way used by nearly all uncivilized people, and any uninstructed per-

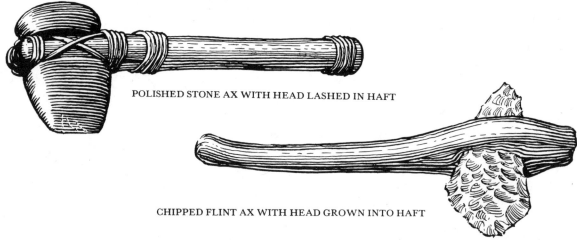

POLISHED STONE AX WITH HEAD LASHED IN HAFT

CHIPPED FLINT AX WITH HEAD GROWN INTO HAFT

Some spears also were made for throwing. These are usually shorter and lighter than regular spears and are called javelins. In Australia the bushmen learned to use a throwing stick to increase the range of their small javelins. It was very accurate in practised hands, and the natives easily outshot the smoothbore muskets of Captain Cook's men in a friendly match. Similar throwing sticks made of reindeer horn were used by the cave men.

An arrow is actually a small javelin and a bow is a better kind of throwing stick. An arrow shot from a bow has great penetrating force. There are records of Indians shooting arrows clear through buffaloes and the English longbow was even more powerful. Some prehistoric bows have been found

son who handles a bow for the first time will instinctively pinch the end of the arrow and pull.

The prehistoric arrowhead was set in a cleft made in the front end of the shaft, and both head and feathers were lashed in place. Making a hole in a flint arrowhead to receive a shaft was not only difficult but impractical, since the head broke when the arrow hit its target. Each of the larger groups of ancient people had its favorite or traditional shape for arrowheads. All arrowheads were generally triangular, but some were quite blunt while others were very pointed; still others were barbed, notched for lashing or plain. So strongly did each "nation" stick to its own pet shape that scientists are able to trace their migrations by the

16

FLINT KNIFE, the edges probaby chipped by pressure

THROWING STICK OF THE AUSTRALIAN BUSHMAN

AMERICAN INDIAN BOW AND ARROW

AMERICAN INDIAN SHOOTING WITH BOW AND ARROW

17

*a* Leaf-Shaped,
Prehistoric European

*b* Triangular,
Prehistoric European

*c* Tanged,
Prehistoric European

*d* American Indian

*e* American Indian

*f* From Swiss
Lake Dwelling

flints they left behind. Contrary to the accepted idea, the average American Indian was not a remarkably good archer. He depended on his wonderful ability at stalking game to get so close to it that he couldn't miss. Buffalo and deer were often shot from a distance of a few feet.

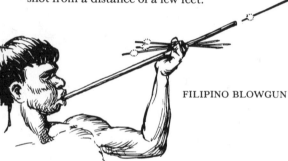

FILIPINO BLOWGUN

A dart with a fluff of fibers at its rear end instead of feathers can be blown through a long tube by lung power and delivered with real force. Such blowguns were used in medieval Europe and are still used in the jungles of Brazil, in the Philippines and on the Malay Peninsula. The darts are much lighter than any arrow and are sometimes tipped with large thorns which not infrequently are poisoned.

There is one astonishing primitive weapon which was developed in only one place in the world—the Australian boomerang. This is a deceptively simple bent stick of precise shape and balance which when thrown follows a curved path, spinning end for end as it travels. An expert can throw it accurately at a mark and it strikes with a vicious blow. It can also be thrown in such a way as to make it return to the thrower. This feature makes it a dangerous gadget for casual experiments.

The Hopi Indians of the Southwest have **curved** throwing sticks which they throw at jack rabbits. These have been called "boomerangs" because they look almost exactly like the Australian weapon, but when they are thrown they act not **like** boomerangs, but like plain, crooked sticks.

## Copper and Bronze

### (5000 B.C.–1000 B.C.)

Copper was the first metal which man learned to work. When he first found it, probably in more or less pure chunks, it was just another kind of stone to him. He knew nothing of purifying it by smelting or of forming it by casting; he just beat it into shape. The fact that it wouldn't break easily and that its shape could be changed without knocking pieces off it made it most desirable.

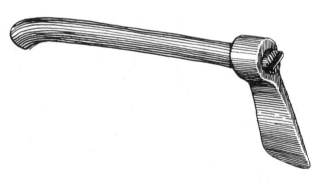

COPPER AX (the handle is imaginary)

AUSTRALIAN BOOMERANG

The use of copper for weapons seems to have begun around 5000 B.C., and for a couple of thousand years it was rare in most places. Around 3000 B.C. a Sumerian must have used a chunk of copper ore to prop up the sticks of his fire, possibly a piece too impure to be shaped by hammering and therefore worthless. Anyway, the heat melted the "stone" and it flowed into a gleaming pool which cooled into a shape entirely different from that of the original piece. Only the copper melted; the heat did not affect the impurities in the ore. Ordinary common sense could see at a glance the possibilities in this.

Later on, somebody smelted some ore which happened to have a little tin in it as a natural impurity. The tin melted along with the copper, and the result astonished the metallurgist by appearing to be a new metal. It was much harder; you couldn't beat it to shape so easily but you could give it a cutting edge which had real authority and a sword made of it wouldn't embarrass you by suddenly folding up in your hand. This was bronze. It has been an important metal for weapons ever since.

CHALDEAN WARRIORS 2000 B.C.

Copper and bronze weapons first appeared in the lower part of the Tigris-Euphrates Valley in what is now called Iraq, and from there spread in all directions. Chaldean warriors in these regions used spears, arrows, battle-axes and maces with copper and bronze heads, and those who carried swords carried bronze ones. The arrows they used seem to have been very long compared to the length of their bows which, in order to accommodate the arrows, actually could be bent into a semicircle. Some experts think this is possible only with a bow made of two separate pieces joined in the middle at a slight angle, and the representations of these bows include this characteristic. Since the only knowledge of them comes from sculptures, it isn't possible to tell how the two halves were fastened together. You can see from the drawing at the left that the relaxed bow hanging on the warrior's shoulder looks as if it were made this way.

Such a weapon would have been a "self-bow," which is one made entirely of one kind of wood. At about this time, or not too much later, the composite bow was invented in this part of the world. Though it isn't very big, it is the most powerful kind of hand bow that has ever been made. Its use spread all over Asia. No doubt composite bows are still used there in remote sections; they were the principal Asiatic weapon until they were displaced by guns.

ASIATIC COMPOSITE BOW

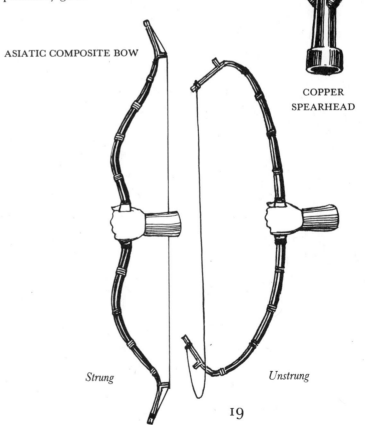

COPPER
SPEARHEAD

*Strung*                    *Unstrung*

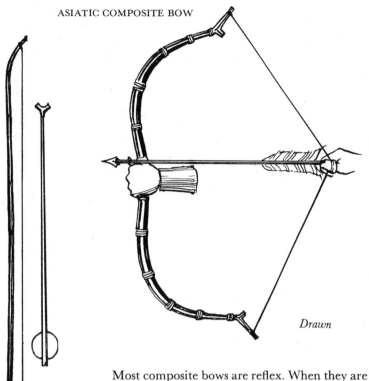

*Drawn*

EGYPTIAN
BOW AND ARROW

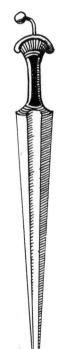

EGYPTIAN SWORD

Most composite bows are reflex. When they are unstrung reflex bows actually bend in the opposite direction to the way they are bent when the string is tight. A composite bow is built up of three different materials: a thin wooden stave, flat on both sides, runs the full length of the weapon, serving as a framework; a layer of split horn is firmly attached to the belly, which is the side toward the archer; when the bow is bent this is compressed but it has the quality of springing back to its original shape. The back of the bow, away from the archer, has dried animal sinews lashed and glued to it; these also return to their original length after being stretched by the bending of the bow. Few composite bows are more than four feet long.

To give you an idea of how good these small bows are: At an archery match in England in 1795 one Mahmoud Effendi, secretary to the Turkish Ambassador, made the English longbows look silly and goggled the eyes of the local archers when he shot an arrow 482 yards with a composite bow. The Turk didn't seem to think his shot was anything much, and mentioned casually that the Turkish Sultan Selim had done nearly twice as well! Mahmoud has been beaten many times in recent years by American longbowmen, but if Selim really shot 972 yards, he still holds the world's record. The American regular-style flight-shot record, made in 1949 by Jack Stewart, is 640 yards.

It's odd that the Egyptians used wooden self-bows because wood was the scarcest material in ancient Egypt. Some of these weapons have been found in tombs. They are straighter than the longbow and don't taper so rapidly; otherwise they are nearly the size and shape of the longbow as it was a couple of thousand years later. Some Egyptian arrows had queer double points, and like most Egyptian weapons seem oddly shaped to us.

The early Chaldeans don't seem to have bothered much with any defensive armor beyond a bronze cap and a shield, but the Assyrians who followed them had body armor of overlapped bronze hoops, as well as chest protectors made of many layers of linen stiffened and glued one upon another. These protectors were also used in Egypt; they were comfortably light and would turn the cut of a sword but not the thrust of a point. Scale armor was also worn by the Assyrians. This was made of many small metal plates sewn onto a leather jacket so that each row of scales overlapped the row below it. Scale armor has the longest history of any; some of Oliver Cromwell's soldiers wore it.

The solid body of foot soldiers, shoulder to shoulder, shield to shield, and bristling with spears, which is known as a phalanx and which we usually associate with the Greeks, was actually invented in Chaldea. We'll take a closer look at the phalanx when we come to Greece.

The Chaldeans also introduced the idea of fighting from chariots. Even before horses were tamed, they hitched small, wild asses to their battle carts and gained a moving, elevated platform of enormous advantage in archery and javelin work. There is little doubt that all the ancients stood in great awe of the horse, and even after he was tamed, possibly about 2500 B.C., he terrified his master almost as much as he did his master's enemy. The writer of the famous description of the war horse in the Book of Job was, to say the least, impressed by his subject: "He paweth in the valley and rejoiceth in his strength: he goeth on to meet the armed men. He mocketh at fear . . . . He saith among the trumpets, Ha, ha!; and he smelleth the battle afar off, the thunder of the captains, and the shouting."

20

EGYPTIAN WAR CHARIOT

Both the Egyptians and the Assyrians mounted two or more quivers on their chariots to hold an extra supply of arrows. A mace or a battle-ax was usually hung on it in a handy position. Not until quite late did anyone fight from horseback and the Egyptians never did care much for it.

When men first invented the village for mutual protection, they undoubtedly surrounded the group of dwellings with a stockade of logs. In time the logs were replaced by stone walls and, as the towns grew and enemies became stronger, the enclosures were made larger and the walls higher. The measurements of the walls of Nineveh, the later capital of Assyria, have come down to us: they were 120 feet high, 30 feet thick and 50 miles long; upon them were set 1500 towers.

The capture of such a place, and the Babylonians finally did capture it, required something more than foot soldiers and chariots. The attackers had to get over the walls or through them or under them. To get over them they may have used ladders or movable wooden towers as later besiegers did; to get under them they probably dug long tunnels or "mines" as the Romans later did; to get through them the Babylonians used a kind of battering-ram in the form of a huge metal-tipped spear mounted on wheels and rammed repeatedly, point first, against the gates by the combined strength of many men.

It isn't known that the Assyrians or Babylonians had any kind of siege weapon which would throw large stones or spears. The earliest mention of such a contraption is in the Bible which says that King Uzziah, who lived in the eighth century B.C., made engines "to be on the towers and upon the bulwarks to shoot arrows and great stones withal."

## The Greeks and Bronze
## (about 1000 B.C.)

The weapons of the early days of Greece were mostly bronze. Though iron was known to them in the time of the Trojan War, it wasn't much used, possibly because it was hard to get but more probably because the smiths hadn't yet learned to work it well or to harden it.

What is known of war and weapons in this early time comes from the *Iliad* of Homer and from excavations like those at Troy and Mycenae. The later Greeks painted the Trojan heroes on their vases, but styles had changed and they missed by a mile the actual appearance of the warriors who sacked Troy.

The two warriors on page 23, each differently equipped, have been reconstructed from earlier

BABYLONIAN BATTERING SPEAR

22

sources. The warrior with the tall, straight-sided "tower" shield is wearing a bucket-shaped bronze helmet decorated with a horsehair plume. On the upper part of his body he wears leather armor; on his feet, sandals; and, on his lower legs, shin guards called *greaves*. That's all the clothes he wore—the rest of him was naked. His spear was his principal fighting weapon; a dagger and perhaps a sword were a second line of defense if the spear was lost or broken.

The Homeric soldier with the figure-eight shield in the shape of a bull fiddle wears a helmet which was at one time a complete mystery. It was described by Homer as being made of boar's tusks, which made no sense to anybody, and no one connected it with wall paintings of an odd-looking helmet showing bands of curved marks until an actual helmet made of parallel rows of hog's teeth was found in a tomb.

The cuirass or breast plate of this warrior was made of overlapped metal bands, a design probably based on the arrangement of human ribs. Instead of greaves, he wore leather leggings and carried a couple of light javelins, one of which he has already thrown.

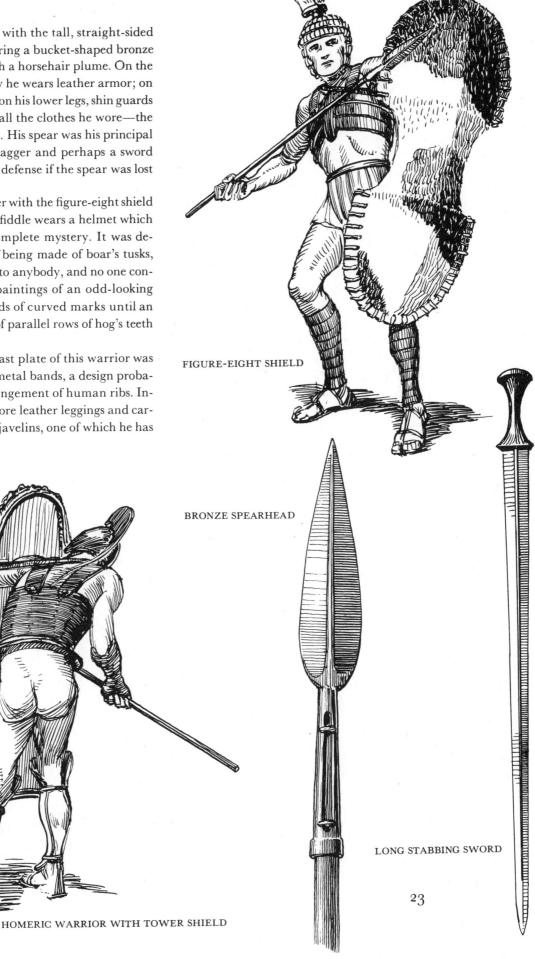

FIGURE-EIGHT SHIELD

BRONZE SPEARHEAD

LONG STABBING SWORD

HOMERIC WARRIOR WITH TOWER SHIELD

23

Both the tower shield and the figure-eight shield had a strap at the top, called a *telemon,* which passed over the wearer's right shoulder. The telemon allowed the shield to be carried on the back in marching and held it up in a fight, so a man could have the use of both arms in a tight spot. Probably the tower shield was made of wood, covered with hide or metal. The fiddle shield (scientists call it a figure-eight shield) was almost certainly made of bull's hide stretched on a frame with the hair left on. The paintings of it, crude as they are, always show a splotched pattern like a bull's hide.

Though there were some pitched battles at this time, much of the fighting seems to have been single combat between armored heroes while the ranks of both armies simply stood and watched. These combats were preceded by long wrangles of the you-and-who-else variety, delivered from behind shields. It's been suggested that the talks served as endurance tests to see which man would tire first from holding up his heavy shield.

In general battle most of the heroes had chariots but usually dismounted from them to fight. Chariots were for transportation on the field and as a means of escape from tough situations. Occasionally a hero would cast javelins from his chariot. No Greek of this period seems to have fought on horseback.

The ordinary Greek foot soldier of Homer's time carried a small round shield which had a simple handhold in the middle of its back; otherwise he wore no armor of any kind. For weapons he had a couple of javelins and a woolen sling, so it isn't surprising that he let the heroes do much of the fighting. The bow was well known to the early Greeks and to hear them tell it, they were good archers, but they used it as a hunting weapon. Only a few specialists occasionally shot an arrow in battle. Ancient Greek drawings and sculptures show both the European wooden bow and the Asiatic composite kind.

If the Greeks tried to break down the walls of Troy, they failed. You've probably read how they are supposed finally to have taken the city: they pretended they'd given up, took to their ships and rowed away, leaving behind them a huge wooden image of a horse. The rather gullible Trojans expended a lot of energy dragging the thing inside their walls. That night, while the town was sleeping off its "victory" celebration, the horse disgorged

HOMERIC FOOT SOLDIER

THE TROJAN HORSE

from its interior enough Greeks to get the city gates open and let in their gang, who had returned from behind the nearest promontory. That did for Troy.

This was surprise, treachery if you like, one of the best of siege weapons, which has succeeded many times between Troy on the Mediterranean and Trenton on the Delaware.

## The Greeks and Iron
## (680 B.C.—C. 150 B.C.)

It isn't possible to say exactly when the Greeks learned to work iron properly. It was a better metal than bronze for swords and spearheads and was used for such weapons, while bronze remained more popular for helmets and body armor.

This section has been dated from 680 B.C. because at about that time there was a change in Greek fighting gear and in the appearance of the Greek warrior. He began to wear a metal breastplate and to carry the big round shield which we think of as Greek. This soldier was the *hoplite,* the elite fighting man of Greece. He came from one of the three upper property-owning classes and was subject to service between the ages of eighteen and fifty-nine.

All Greek citizens were subject to military draft. Those below the third property class served as *pel-*

GREEK HOPLITE, ABOUT 500 B.C.

25

*tasts* or javelin men. The peltasts were equipped with slings as well as javelins and carried small, round shields as their Homeric forebears had done, but they were now protected by helmets, greaves and leather shirts.

The famous Greek phalanx was made up of hoplites. They stood in solid ranks with their shields overlapping and their spears thrust forward. Since those spears were about twenty-one feet long, even those carried by the sixth rank projected well ahead of the front line of shields and the enemy was faced by a very prickly affair. The earliest phalanx had about two hundred men in it. By the time of the Persian War these had been increased to about five thousand; still later, when Greece fought Rome, there were sixteen thousand men in each phalanx.

Local war was polite and quite formal at this time. It wasn't considered good form to attack armies on the march, or encamped for the night, so there were no scouts and no guards. Each hoplite had at least one servant with him and had to provide his own food. Greece isn't a very large place and the enemy was seldom more than a day's march away, so not too much had to be carried.

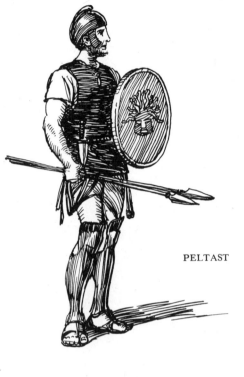

PELTAST

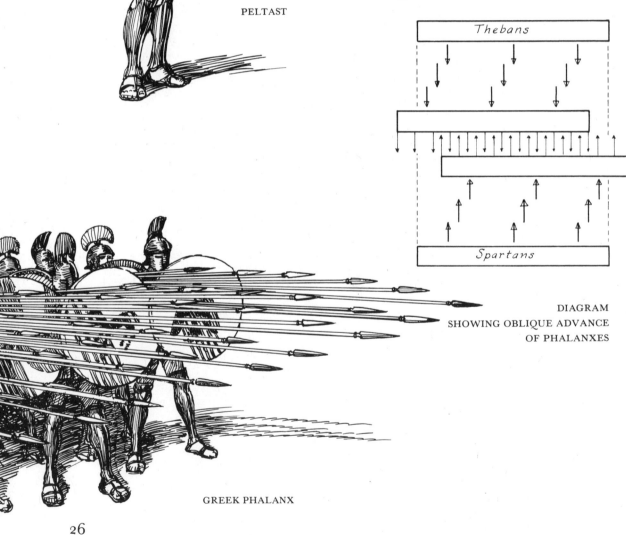

DIAGRAM
SHOWING OBLIQUE ADVANCE
OF PHALANXES

GREEK PHALANX

When the armies met, almost by appointment, on a nice, flat field, they drew up their phalanxes about two hundred yards apart and charged one another at the double. This produced a curious result: since each man's shield protected only his left side, he automatically made a quarter turn to the right as he advanced. Thus the two phalanxes didn't meet head-on as they were originally aimed. Instead, the left of each phalanx was enveloped or outflanked by the right of the other. Epaminondas of Thebes, the first general who deliberately took advantage of this situation, beat the tar out of the Spartans who considered themselves practically invincible. In ordinary neighborly scraps, however, one side or the other would presently ask permission to bury its dead. This was an admission of defeat and everybody went home. Don't think from this that the Greeks couldn't fight; they could and did.

When cavalry began to be used in Greece after the Persian War, the "knights" of which it was composed were drawn from the highest class only. They were equipped exactly like the hoplites except that they wore spurs and carried no shields. Since stirrups hadn't been invented, it was no cinch for a man in armor to mount a horse. Some horses were trained to kneel for the purpose but the really agile knights used their spears to pole-vault aboard! Until the time of Philip of Macedon, 359 B.C., Greek cavalry was more showy than effective.

Like everything else the Greeks made, their weapons were beautiful. Their swords were fairly long and double-edged for slashing. Both edges were curved so that the blade had a narrow "waist" a little below the handle. Most of their spears had handsome leaf-shaped heads, and the javelins had a strap attached to the middle of the shaft to help in throwing.

There were two main kinds of helmet; one a bronze bucket with eyeholes and a nose guard, the other somewhat bowl-shaped with a pierced projection in front which covered the whole face in combat; when this helmet was pushed back on the head the face-piece stuck out in front like a cap brim. High crests plumed with horsehair were sometimes worn on both kinds.

The usual cuirass or body armor was made of two plates, front and back, laced together and connected over the shoulders by curved metal plates.

Some breastplates were molded to the exact shape of the body. After 500 B.C. a fringe of leather tabs was hung from the bottom of the cuirass to protect the belly. A short shirt was worn under the armor and a few inches of it hung below the leather fringe.

Greaves were individually tailored of bronze and fitted the leg so well that they needed no straps to hold them in place. Their fronts were made high to protect the kneecap.

The Greeks are credited with the invention of most of the stone-and-spear-throwing siege weap-

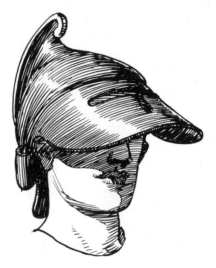

CORINTHIAN HELMET, with visor covering the whole face

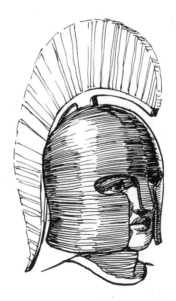

BOETIAN HELMET—THE MOST POPULAR SHAPE

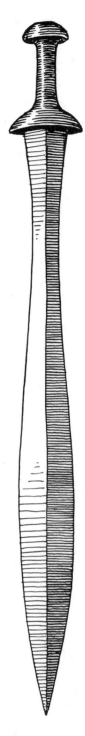

GREEK SWORD

27

ons which were in use until guns became powerful enough to take over; in fact, some authorities think that the Greeks brought projectile throwers to the highest point they ever reached. This may be true, but there is scant evidence today of how the Greek engines worked or what they looked like, so we shall have to fall back on the later Roman ones of which we have some knowledge.

You will have heard of "Greek Fire." This was a much later invention, unknown to the ancient Greeks, but it is dealt with here because it is called "Greek." It was developed in Byzantium long after the fall of Rome and was used to frighten the daylights out of the Crusaders. Greek Fire was poured from city walls upon the heads of besiegers or thrown in bottles by hand and by "engine"; and there seems to have been some way of blowing it out of a tube. It was the napalm* of its time; it clung to whatever it hit, burned fiercely and was believed to be inextinguishable.

The exact composition of Greek Fire was top secret in its day and it has remained so. It probably contained pitch, resin, grease, powdered metal and some form of petroleum, possibly naphtha. Its mystery, and hence its ability to scare an enemy, increased its value as a weapon far beyond the real damage it did.

## The Roman Soldier
### (200 B.C.—400 A.D.)

As the company is the basis of the modern regiment, so the century was the basis of the Roman legion. A century was commanded by a centurion, with a standard-bearer as second in command. Two centuries made a maniple, three maniples a cohort, ten cohorts a legion. A century wasn't invariably a hundred men; it could be more or less, and the size of a legion varied at different times from as few as three thousand to as many as six thousand men.

The legionary was a heavy foot soldier. Although his weapons and armor altered somewhat in the course of Rome's long history, the changes were not great and across five hundred years the un-

* Jellied gasoline used in flame throwers and incendiary bombs.

compromising edge of his short sword was felt from Egypt to Britain.

The original design for the legionary's hooped cuirass was borrowed from the Greeks but the Romans tailored it to their own fancy. The hoops which encircled the chest were hinged at the back and clasped in front. They were actually supported by the leather tunic to which they were sewn. The shoulder pieces, made in four strips, were less cumbersome than the single plate of the Greeks. The leather tabs at the bottom of the cuirass were retained by the Romans, and they added others over the upper arm.

The Roman helmet also was copied from a Greek style but it was much less fancy and more practical than the original. Reinforcing bars crossed one another at the crown of the head and at the crossing there was a ring to support the crest. Most Roman helmets had hinged cheek pieces and a small bar across the front as a visor. On the march the helmet was carried on the right shoulder.

The legionary's shield was oblong with or without cut-out corners. On its slightly bulging face was the insignia and number of the legion to which its owner belonged. Its height was supposed to be the length of a man's arm. In later days when the Roman soldier was a hired man without the old punch, his shield became much larger and oval in shape; and the famous short sword was lengthened, to keep the enemy further away.

In its heyday this sword was about twenty-two inches long, double-edged and perfectly straight; the point a quite obtuse angle. It was worn on the right side hanging usually from the kind of belt which is called a baldric, passing over the left shoulder. Spain was sword-maker to Rome and she kept her fame for fine blades for centuries after her first customer had passed on.

More important to a legionary than his sword was his *pilum*, the heavy spear with which he wrote history. The legionary on page 29 holds one in his right hand. It was only about five and a half feet long but at least a third of its length was iron head. The thick wooden socket took another third in the middle. This was used to ward off blows and was swelled a little where it met the handle, so it could serve as a hand guard. The pilum served as a regular spear but it was also thrown as a javelin.

LEGIONARY HELMET

ROMAN CENTURION AND LEGIONARY

SWORD

29

In a Greek phalanx each soldier was allowed a total of three feet of space, and he could act only on orders and in unison with the rest of his gang. A legionary in ranks had three feet clear on each side of him and what was much more important, he was allowed to use his head and did so. When the Romans met the Greeks, drawn up in solid array on a nice level pasture, they laughed, marched around them and took the town the phalanx had supposed it was guarding.

When legion met phalanx on rough ground the battle tactics were changed. Charging with locked shields on a boulder-strewn hillside presented difficulties. The legionaries ducked under the wavering shields and spears, thrust upward with the pilum and Greece became a Roman province.

Not all of Rome's soldiers were legionaries. There was cavalry called *equites* composed exclusively of blue-bloods. They ordinarily wore scale armor and carried a small oval shield and a spear lighter and longer than the pilum.

There were also light-armed foot soldiers called *velites*. These wore iron skull caps and leather tunics, sometimes studded with metal in the style called jazerant. The velite shield was oval and about two feet high. Though they were equipped with swords and light spears, the velites were primarily sling-men. Their lead sling-pellets were specially cast and had "thunderbolts" on them or some bright remark like "Take this!" The same idea has been expressed in chalk on modern artillery shells. The Romans made good use of the mobility of the velites in carrying out end runs.

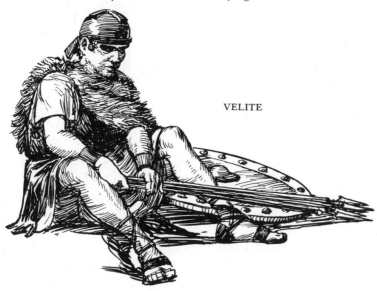

VELITE

High officers wore solid bronze body armor molded closely to the shape of the body. In early days the breast piece extended down over the belly and usually had some kind of ornamentation which was raised a little from its surface. The leather fringe worn with this armor was often plated with metal. Officers wore a cloth tunic under their armor and an official military mantle over it, knotted on the right shoulder. Their helmets, swords and shields were much like those of the legions but fancier. Oddly, no Roman sculpture shows an emperor or a general wearing a helmet.

GLADIATOR'S HELMET

The gladiators who entertained the Roman citizenry by slaughtering one another in public wore special armor for their work, and also used some special weapons; for instance, one group of specialists used a *net* and a *three-pronged fork!* The first gladiators were captive slaves but later freemen went in for gladiating and attended schools to learn the business.

## Roman Sieges and Siege Engines
(200 B.C.—400 A.D.)

The Romans were experts at both fortification and siege. Rome itself was surrounded by walls which were increased in length three times to accommodate the growth of the city; and in Roman Britain the northern frontier was guarded by a sixteen-foot wall built seventy miles long, clear across the island. Of more interest here are the Roman methods for getting past other people's walls.

When they besieged a town the Romans surrounded it with a log stockade or an earthwork, built just beyond bowshot. Sometimes, for a long siege, the walls were built double with a roof over the space between them. The object, of course, was to make sure that no one escaped from the city and above all that no food was brought into it. Since methods of preserving food at that time left much to be desired, not too large a supply could be stored and a besieged city was usually cleared of dogs, cats and even rats by the time it gave in; and in the end, it almost always gave in.

The oldest siege weapon, excepting the Assyrian spear-on-wheels which was really the same thing, was the battering ram. In its Roman form this was a whole tree trunk tipped at its business end with a heavy iron ram's head. The log was suspended from a framework by ropes which allowed it to be swung forward and back. The swinging was done by as many as fifteen hundred men, the nearest ones working under a heavy roof called a testudo or tortoise, which took the brunt of the boiling oil, molten lead and other unpleasant substances which the besieged dumped from their battlements.

Mining was nearly always tried, the attackers tunneling from behind their wall and attempting to pass under the wall of the city. Inside the town on quiet nights the officer of the watch would lay his shield on the ground hollow side down and put his ear to the top of its bulge. If a mine was advancing, it would reach a point where the officer could hear the diggers. When he located them the town's next move would be a counter-mine. Usually this was a deep trench into which the sappers would unexpectedly break. If plenty of water was available they could then be drowned in their tunnel; if not, they were allowed to dig themselves into an oil-soaked brush heap which was set afire in their faces.

Sometimes, working at night, the besieger built an earthen ramp from his wall towards the city wall, a little closer each night; but if the townsfolk were on their toes, they built *their* wall facing the ramp a little higher each night, as well as doing their best to make life short or at least uncomfortable for the ramp-builders. The next move of the invaders would be to push a fighting tower out onto the ramp and attack from it with throwing engines and arrows, finally lowering a drawbridge from its top to the town wall and meeting the defenders hand to hand.

During all these operations there was a constant

SIEGE WITH BATTERING RAM, TESTUDO AND
FIGHTING TOWER

31

two-way bombardment of big stones thrown by ballistas, onagers and catapults; heavy, medium and light artillery respectively. Basically all three worked the same way: a very thick skein of cords was twisted to a terrific strain which was suddenly released upon a projectile. It isn't known just what kind of cords were used; they stood up under weeks of constant use, and hair or animal sinew seem the likeliest substances. Nothing modern experimenters have tried will stand the gaff for more than a few shots.

The onager was the simplest of the engines because it had only a single horizontal skein with one vertical beam inserted in it. The skein was twisted tight by geared winches, working on both its ends. This was done with the beam upright. To load, four or more men manned a windlass and pulled the pole back and down until it was nearly horizontal and had put a still greater twist on the skein.

Onager means wild ass. It earned the nickname from the soldiers because its rear end kicked up when it was discharged. The onager was usually mounted on wheels. Some onagers had a kind of scoop at the beam's end to hold a stone for throwing, but on the more effective ones the stone was put into a sling. One side of the sling was attached to the beam, the other side merely being hung on a pin at the end of the pole. This slipped off when the beam was part way up and the stone sailed free. Using a sling added about a third to the distance a stone could be thrown. With either kind the beam was restrained by a slip-hook which could be tripped instantly by a yank on its lanyard.

The heaviest and the lightest Roman artillery were very much alike in basic design and probably were built in all sizes from a monster four-ton ballista down to a hand catapult which had no more weight or power than a medieval crossbow. The heaviest ballista, once it was aimed, continued to throw its rocks at the same spot time after time. It could heave a sixty-pound rock as far as five hundred yards. This was pretty good. An Amer-

ONAGER IN ACTION

ONAGER LOADED SHOWING SLIP-HOOK

ican Revolutionary naval cannon had little more than twice that range and little better accuracy with a shot weighing half as much. A ballista could outrange an onager. So, while the "wild ass" was lobbing headsize rocks just over the walls, the ballista was plunking boulders well into the center of the town.

You may read that the ballista was an oversize crossbow and in medieval times this was true. In those days the ballista was a lighter job and its force came from the bending of a large wooden bow; but the Roman ballista was powered by torsion, like the "wild ass." It was the construction that was different. Instead of a single beam operating vertically, the ballista had two short arms which moved horizontally, each arm with its own separate, vertical skein of cords. Force was transferred to the projectile by a heavy bowstring which connected the ends of the arms. The windlass which drew the "bow" thus formed, didn't pull directly on the bowstring but was hitched to a sliding trough in which the ammunition was placed. The bowstring was restrained by a trigger mechanism which was fastened to the trough. Pawls on the sides of the trough engaged teeth on the frame and prevented the trough from moving forward with the projectile when the trigger was sprung.

Catapults were really small ballistas, but they were swiveled so that they could be aimed readily from side to side, and they were balanced on a pin so that their elevation could easily be changed. This suggests that they made some pretense to accuracy. A short, heavy javelin was their usual ammunition. The catapult bowstring was held by a double-trigger hook fastened on the rear end of the trough and clamping the string on both sides of the javelin butt. Springing the trigger simply allowed the hook to rise, and the released bowstring whanged the dart forward in the trough to a good start in the right direction. A large catapult had four foot arms and threw a six-pound javelin some five hundred yards.

BALLISTA

CATAPULT FOR JAVELINS. The legionary is dressed for winter

PART OF A CATAPULT SHOWING TRIGGER OPERATION

# The Dark Ages

Rome was the keystone which sustained the European civilization of her time and when she fell the whole structure went down with her. So complete was the demoralization that even records of it are fragmentary. It seems incredible that in so short a time men could forget all that had been learned but that is exactly what they did; and it was fear that caused them to do it. Law disap-

peared and with it all security of life or property. For mutual protection men huddled in little groups under some strong leader. Sometimes they hid in a former Roman stronghold and defended it as best they could or used it as a base for raids on their neighbors.

It was out of the welter of petty chieftains, struggling first for survival, then for supremacy, that feudalism was born: each man receiving protection from and giving service to another more powerful than himself. The earliest massing of these scattered groups in any real strength was that of the Franks under Clovis, about 480 A.D.

EARLY FRANKISH SHIELD

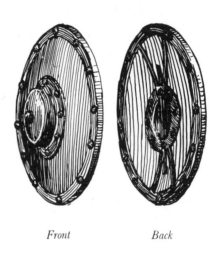

*Front*  *Back*

FRANKISH WARRIOR, SIXTH CENTURY

FRANKISH THROWING AX

As the result of tomb probing some notion has been gained of the equipment of Frankish warriors. They used iron and shaped it well but they hadn't learned to harden it. Their spears were iron-tipped; their thirty-inch swords were iron, but poor things; iron rimmed and braced was their round wooden shield with a large iron boss mounted in its center. Iron-headed too was the *francisc,* the curiously shaped throwing ax which was the Frank's prime weapon.

The Frankish warrior used no armor except a leather cap reinforced with crossed metal bands. He wrapped his legs to the knee with strips of cloth or leather over the long "trews" or pants he wore. On his body was a belted fur jerkin reaching half-way to his knees and giving him some protection.

In England the Anglo-Saxons used very similar

35

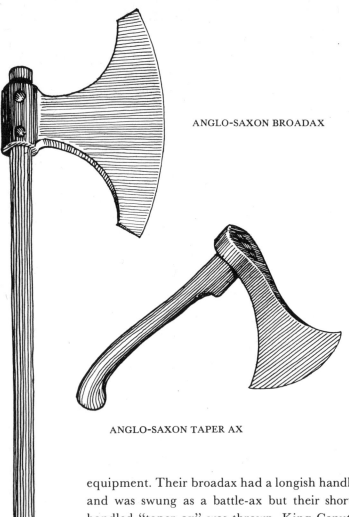

ANGLO-SAXON BROADAX

ANGLO-SAXON TAPER AX

This was the first of the "pole arms," and was used as long as any of the pole arms. Only a few examples of English military bills still exist, because it seemed only sensible to the ex-soldier, returning to the farm, to put his bill back to its original work, and most of them were worn out that way. The Saxons also worked a variation on the mace which must have had great possibilities in the hands of a good man. It was called by a gentle and poetic name—the "morning-star." Its heavy, usually spiked head was attached to a handle by a short length of chain, and though it might be a little hard to control, when it did land it took the fight out of the toughest Norman.

The beginnings of the system of vassalage and knighthood were set up in England about the time of King Alfred the Great (872). King Arthur, if he lived at all, is assigned to a period about two centuries earlier. That would make him a half-wild chieftain. His Round Table of knights was invented for him by later romantic legends. The paintings of them in fourteenth-century plate armor hit some kind of high mark for the ridiculous. Actually, only a few wealthy Saxon leaders could afford chain

equipment. Their broadax had a longish handle and was swung as a battle-ax but their short-handled "taper ax" was thrown. King Canute measured some land by marking it "as far as a taper ax can be thrown."

By the time of Charlemagne (c. 800) the Franks, though clinging to the francisc, had learned to harden iron; and the *lorica,* a jerkin of chain mail, began to be worn by him and his men. Charlemagne's sword was longer than Clovis's, and it had a crossed guard at the hilt which was used on swords for six hundred years after him. Some of the crack troops of the Franks now wore helmets with scalloped leather curtains which hung about their faces, and they ornamented their legs by crisscrossing the wrappings all the way up. This was a general style almost everywhere at this time and is the ancient basis of the design of the Scottish Argyll socks.

The Anglo-Saxons had discovered that their long-handled pruning bill was useful for lopping off limbs other than those which grew on trees.

FRANKISH SOLDIER, NINTH CENTURY

mail or jazerant jackets and iron hats; most of them fought bareheaded in their shirts.

The Anglo-Saxons used the bow, but chiefly as a hunting weapon. In war they depended more on the sling, a Roman habit which they may have adopted during the Roman occupation of Britain.

Once the cavalry of mounted knights was established, the common foot soldier became and remained for some centuries almost useless. He was armed with whatever he could pick up around home, and he could rarely do any real damage to a mounted man in chain mail. Some yeomen were used as slingers and archers and some few had arms given to them, but for the most part they seem to have impeded the knights who did most of the fighting, as much as they helped them.

The Dark Ages remembered nothing of the Roman science of fortification. They began again with wooden stockades and earthworks. By the beginning of the Middle Ages, men were building stockades, ditches and drawbridges for defense and with these, at least as early as 585, came a return of the battering ram and some kind of ballista. The old "tortoise" to protect the men at the ram also came back with the new name of "snayle."

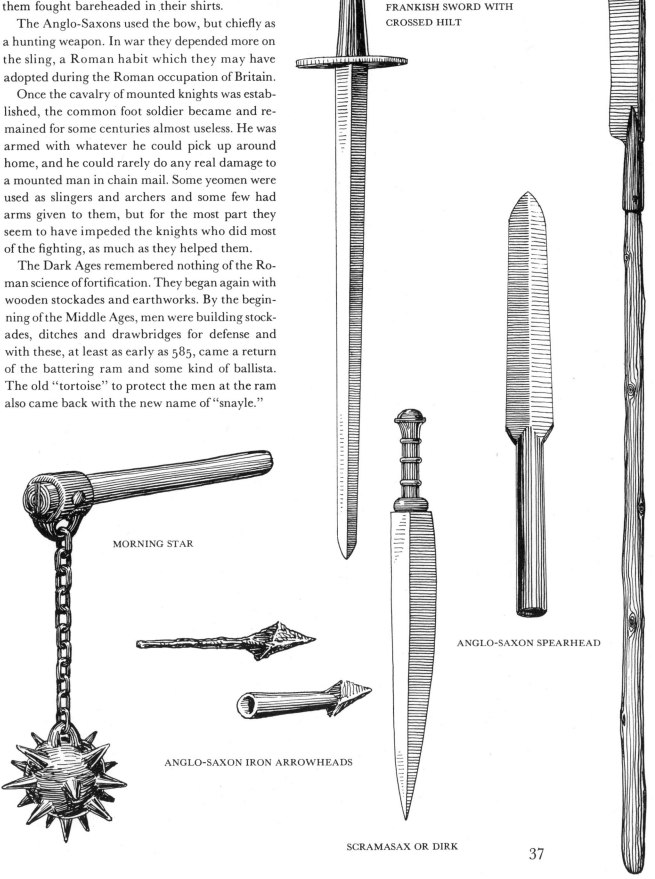

FRANKISH SWORD WITH CROSSED HILT

MORNING STAR

ANGLO-SAXON IRON ARROWHEADS

ANGLO-SAXON SPEARHEAD

SCRAMASAX OR DIRK

ANGLO-SAXON BILL

37

## The Norman Conquest (1066)

The Bayeux Tapestry which illustrates the whole story of the invasion of England by Duke William, and depicts the Battle of Hastings which King Harold and his Saxons lost to the invader, shows the doings of knights and men-at-arms but largely disregards plain soldiers.

Much of what is known of military equipment at this time comes from the Tapestry, though not all the information it provides is wholly dependable. For instance, it is known from other sources that the Saxons wore kilts, but the Tapestry shows them dressed like the Normans in divided knee-length hauberks of chain mail or scale armor. The hauberk was descended from Charlemagne's lorica. In front and behind it was split to the crotch, to allow its wearer to ride a horse. Hauberks had short sleeves and most of them had hoods with an opening for the face. They were usually topped

TWO SAXONS AND
A MOUNTED NORMAN KNIGHT

38

with a conical iron cap which had a nose guard attached to it.

A quilted jerkin was probably worn under the hauberk, as it always was later, and the warrior's legs wore banded-over narrow trews, such as those worn in Charlemagne's time. These trews, also, may have been leather. Some important people seem to have worn chain-mail leggings.

Norman shields were usually round at the top and pointed at the bottom and from three to four feet high. Nearly all of them had painted decorations on them but none of the patterns were associated personally with the bearer, as they were in later heraldry.

Each mounted knight carried a wooden lance with an untapered shaft probably eight or nine feet long and tipped with a broad iron head. Some "couched" the lance under the right arm in the new fashion which had become possible with the introduction of stirrups, but more thrust with the older overarm stroke. The lance was frequently thrown like a javelin and its butt rested on the stirrup when it wasn't needed. Swords had by this time reached their full growth; broad near the hilt and tapered to the point, their double-edged length was a full forty-four inches from pommel to tip. This was the "knightly blade" which, changing very little, "carved the casques of men" for nearly four hundred years. Its ornamented scabbard hung from a belt straight down at the knight's left side. At his saddle bow the Norman warrior carried either an iron-headed mace or a broad-bladed battle-ax according to his taste.

King Harold was killed by an arrow at Hastings and the Tapestry shows massed bowmen fighting for the Normans. On the Saxon side there is only an occasional archer mixed in with men-at-arms, who fight with bills, spears and axes. Thus the bow appeared on both sides in this ancient battle, but though the Conqueror's son Henry encouraged archery by ruling that accidental shootings at practice shouldn't be punished as crimes, the bow as a major English war weapon didn't come into its own until many years had passed.

There had been an elaborate feudal system among the Saxons, but the Normans swept most of its regulations aside and set up their own. Among them military rank was a thing wholly separate

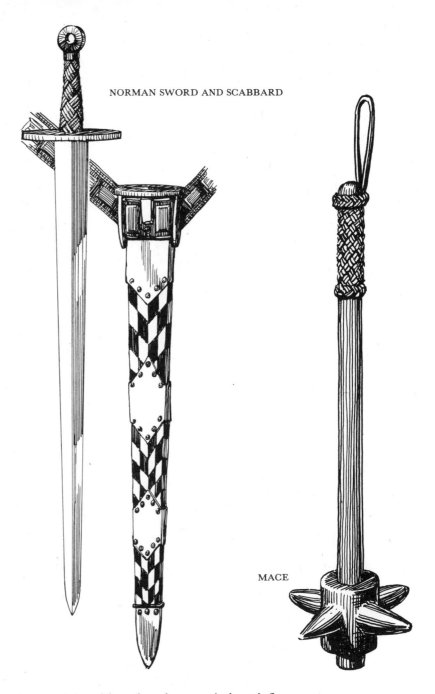

NORMAN SWORD AND SCABBARD

MACE

from social position, though not entirely uninfluenced by wealth. The king's son might be a mere squire and a man-at-arms might aspire to become a knight, as a result of some remarkable feat in battle. The next step higher, however, the knight banneret had to "own" and equip at least fifty men-at-arms; and to do that took cash or its equivalent, which was land. He who held land must defend land. The usual military service was forty days. Clergymen and ladies of estate were not required to serve personally but had to furnish and equip substitutes or pay *scutage*. This was

39

"shield money" which in easy times any vassal could pay to the king for release from the obligation of military service; the king could use the money to hire mercenaries who were always available; in fact, mercenaries fought among Duke William's invasion forces.

It should be understood that a man-at-arms was a superior soldier, potentially, at least, a knight. His lord might pay for the arms of such a one if he was particularly good with sword or spear; but the common infantry had to arm itself as best it could. Armor was far too expensive for ordinary people to own, especially in France, since the common people actually were little more than slaves. The value set upon armor is indicated by the little figures in the border of the Bayeux Tapestry, who are busily stripping mail shirts from the fallen while the battle still rages above them.

It is probably true that few who wore the golden spurs of chivalry actually lived up to all its high ideals of bravery, piety, generosity and purity well enough to qualify as Chaucer's "very gentil parfait knight," but at least those ideals were set up and served to smooth a little bit the rough manners of the time. There were few people then who could read, and the civilizing effect of literature was slight; so even if chivalry later reached fantastic heights of pretense and absurdity, in its early days it shared with religion the job of turning savages into gentlemen.

The aspirant to knighthood normally began his training at about the age of twelve by serving as a page in the castle of a nobleman. During this period he was much in the company of women but he underwent constant training in the use of arms and in horsemanship. At sixteen or thereabouts he became a squire or shield-bearer. In early days he was just that: he rode behind some knight and carried his shield for him. Later, *squire* or *esquire* was an honorable title just below knighthood, which many men bore all their lives.

Unless he greatly distinguished himself before that age, a squire couldn't become a knight until he was twenty. Then, having confessed, he went through an elaborate ritual of fasting, and after an all-night church vigil, he took a great oath to tell the truth and protect all that was

KNIGHTS' SPURS, NORMAN

*Eleventh–Thirteenth Centuries*

*Fourteenth Century*

*Fifteenth Century*

weak, good or holy. When the oath had been sworn, the king or some great lord hit the novice on the neck with the flat of a sword; gilded spurs were then fastened to his heels and he was a knight "without fear and without reproach."

Unfortunately it didn't always stick. Some disgraced themselves so completely that they were stripped of the honor in another ceremony, which consisted of having their spurs hacked off by the king's cook.

ONE PATTERN OF CHAIN MAIL

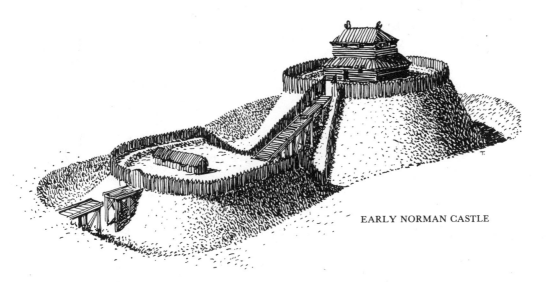

EARLY NORMAN CASTLE

## Castles

William divided England among his leaders and told each man to hang on to what he was given. The Saxons showed a distaste for Normans which suggested to each Norman knight the wisdom of building a stronghold where a night's sleep could be had safely. For the building of it the Saxons were "persuaded" to contribute their labor gratis.

These first Norman structures had but slight resemblance to anything you think of when you see the word "castle." Most of their defenses were earthworks, and such buildings as were put on them were entirely of wood. Partly, of course, this was through haste to get something up (one of them was built in eight days); also, wooden castles were quite usual in Normandy at the time. The Millennium, the year 1000 A.D., was not long past, and since all Christendom had fully expected the world to end then, they hadn't bothered to build permanent structures and soon they had forgotten how to do it. Only a few of the earliest Norman buildings in England were stone, but the conquerors were terrible masons: Winchester Cathedral tower actually collapsed fifteen years after they built it! In trying to make up with sheer bulk for their lack of skill, they built the walls of the keep of the Tower of London fifteen feet thick.

There seem to have been no Saxon castles, because they had never been needed; the Saxons fortified towns. A castle is a private fort. It was invented in France in the Dark Ages and brought over to England.

The Norman began his castle by digging a deep and wide ditch around a circle perhaps two hundred feet in diameter. The dirt from the ditch was thrown toward the center to form a flat-topped mound ringed with an earthen rampart. On the top of the bank a strong wooden stockade was built, and in the center of the ring there was a well-constructed but nearly windowless wooden house. Here the knight, his family, his servants and his men-at-arms, surrounded by their horses and dogs, lived a crude and pungent existence. The entrance to the mound, or *motte* as it was called, was through a single gate reached by a wooden bridge which sloped upward across the ditch. This was literally a "draw bridge" because the planks of one section could be drawn back towards the castle at night, leaving a gap difficult to pass. All other movable bridges were afterwards called drawbridges, no matter how they worked.

The need of more space, especially for the animals, soon led the knight to enclose another and larger ditch-and-stockade area next to his stronghold and surrounding the approach to his bridge. This forecourt was called the bailey; its stockade

41

crossed the main ditch and marched up the mound to join the motte stockade on both sides of the gate. The outer entrance to the castle was now across another bridge leading into the bailey.

The bailey could be defended for a while and then abandoned if necessary, for a last stand on the motte when the knight "burnt his bridges behind him." As time went on the ditches were made deeper and more dirt was added to the motte which made it higher; a few still stand which are as high as a hundred feet. This added height, sloping always towards the center, had the further effect of making the flat top much smaller. Then the wooden house was replaced by a stone stronghold which could itself be defended by a few men. This was the *donjon* or *keep,* and for a while it was still surrounded by the old wooden fence. In time of siege it was necessary to hang wet hides on the timbers to protect them from fire.

The first stone donjons were nearly square and weren't as narrow or tall as they later became. The whole structure was above ground and the first floor, which was used for storage, had no outside door. The entrance to the keep was on the second floor through a small projecting structure, to which an outside stair led upward along one wall. Sometimes the stairs had a gate at the bottom and a drawbridge at the top.

On the inside, the second floor of the donjon was divided into two rooms. One of these was used as living-and-sleeping quarters for the retainers. Here all cooking was done in a corner near the well, the smoke of the kitchen fire finding its way out through a hole in the roof if the wind was right; the other room was smaller and served as a private chamber for the lord and his lady. And the only access to the storeroom was from this room, by a little stair built within the thickness of the wall. Thus his lordship could guard his provisions personally. A similar stair gave access to the top of the donjon wall, which was built higher than the roof of the place and where watch was kept at all times.

In the course of time the wooden fence around the motte was replaced with stone and soon the one around the bailey was also, but the main defense of the castle was still the keep. Its windows were small and very high as a defense against projectiles. New ones were made higher and narrower, like a tower, by the device of putting the knight's chamber *above* the great hall, on the third floor. Sometimes there was an entry floor below the hall, making the whole tower four stories high.

The best assault against a keep was mining, usually under one corner. The thick walls were actually two walls with loose rubble between

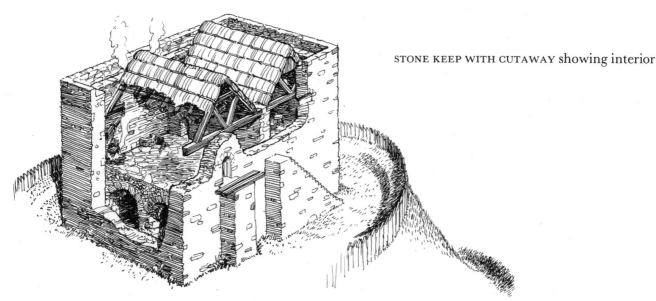

STONE KEEP WITH CUTAWAY showing interior

42

them, so when the mine caved in, the whole corner of the donjon came rumbling down. The wreckage made a ramp for the invader and the place became indefensible. To make mining as difficult as possible the later keeps were built with an extended base, called a plinth. Tower keeps had gaps, called crenels, in the tops of their walls, from which an archer could shoot with some protection from the "merlons" which were left between them. Vertical slots were cut in the walls at lower points, where a man with a crossbow might be stationed. Such a tower, well provisioned and with a few stuffed dummies for show, could be defended by twenty men unless they got dysentery, which they usually did get because sanitation wasn't part of the plan for defending a castle.

TOWER KEEP

CROSSBOWMAN AT A LOOPHOLE

At the time of the Conquest the Norman bow was probably about five feet long, and to discharge an arrow, it was drawn to the archer's chest. At some time around 1100 the bow was lengthened to six feet or more, and the draw was then to the "ear" (actually to the angle of the jawbone); the shooting range was increased

ARCHER SHOOTING THROUGH A CRENEL

43

HOARDINGS

MACHICOLATIONS

greatly. Three-foot arrows began arriving through the keep's high windows and disturbing his lordship at meals. To hold the bowmen out of range, the low walls of the bailey were extended and strengthened and more vigorously defended.

Then, towards the end of the twelfth century, stone-throwing siege weapons began to get better; not only could they clear the outer walls with ease but, worse, the stones hit the tops of the walls and fractured into murderous fragments. The obvious cure was to make the walls higher. This was done and the castle began to take on a proper romantic look.

The main defense of the place was now the outer walls. The keep lost its importance and no new mottes were built. As the years passed (we're covering a lot of time now), the keep was replaced by a strong, square gatehouse with towers at its corners and living quarters were built against the inner face of the high "curtain" walls. These high walls were a good defense against projectiles, but it was hard to keep miners away from their bases, so something more had to be added. The first addition was flanking towers. These were usually round and projected outward so that the defenders could shoot along the face of the adjacent wall. It was soon discovered that the old curved shape of the bailey required too many towers, so castles were then built square.

Next *hoardings* were applied. These were wooden balconies projecting from the tops of the walls, allowing the defenders to drop discouraging substances on the heads of storming parties. Much later the overhangs were built of stone as part of the walls and rejoiced in the thumping name of *machicolations*.

Still more defense was needed. Sappers could start work at some distance and run their galleries underground to the bases of the walls undetected, or at least undeterred. Water was the answer to that; and the wet-ditch, or moat, came into existence. Any tunnel that could be built in those days would be flooded if it were dug under a moat.

Water was kept in the shallow moat around the castle by damming a natural stream. The dam was often separately fortified and guarded to prevent the draining of the moat. The gate on the enemy's side of the moat was called the *barbican*. The drawbridge was lowered from the *list gate*. It admitted to the *list* which was the grassy strip between the walls. Here tournaments and spear practice were held, and, in time, any ground laid out for a tournament came to be called a list.

When the old keep was abandoned, the main gatehouse became the place for a last stand, and the lord and his family lived there in times of emergency. The small turrets on the inner corners

44

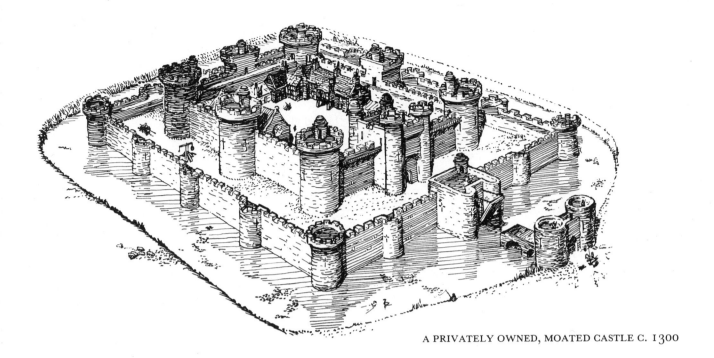

A PRIVATELY OWNED, MOATED CASTLE C. 1300

of the gatehouse and on the larger round towers covered narrow spiral stairs.

With the castle completely surrounded by water (some of them sat in the middle of artificial lakes), the approach was by a bridge so built that a section nearest the castle could be up-ended by chains and windlasses, leaving an impassable gap. The gate defenses were elaborated. Little subforts called barbicans were built on the shore side of the bridge for preliminary defense. On the castle side, beyond the drawbridge, the narrow passage through the gate tower was defended by means of loopholes in its sides and round holes in its ceiling, through which boiling oil and hot pitch could be poured down on an uninvited guest. Exit from the passage to the inner court was barred by a stout oaken gate which could be reached only by passing one or two portcullises. A portcullis was a heavy wood or iron grating spiked along its lower edge. It could be dropped suddenly across the passage from the ceiling.

The foregoing improvements were developed gradually and some of them were added to the older castles, but not until the end of the thirteenth century was a castle built which had all of the defenses and which was for that time, practically impregnable. Almost the only way to capture such a castle was to make a deal beforehand with a pal on the inside to come down at midnight and open the little back gate for you. There was always this little postern gate for sneaking couriers in and out. This was one of the few periods in warfare when the defense had the advantage of the offense.

PORTCULLIS, CLOSED

45

## War Games (1200–1300)

In our day the first interest of most men is business and after that sports; in medieval times the dominant interest was war and after that sports, but there was not too much difference since sports were warlike too. Nearly all of them were some form of combat or a contest of skill which prepared the players for combat. As now, those who didn't enter the game personally thronged to watch.

The most spectacular and exciting of all contests was the tournament or, as it was often called at the time, the "hastilude," which meant a game of spear-play. The first tournament was held in France the same year the Normans invaded England. The idea was enthusiastically imitated and, though often condemned by kings and priests, tournaments flourished all through the Middle Ages. Weak kings, fearful of the power of their nobles, were against tournaments; strong kings encouraged them and even rode in them.

The earliest of these entertainments were not child's play. The knights armed themselves exactly as if for war and went at one another with everything they had, no rules and no holds barred. Considerable blood was spilled, and occasionally one side or the other would talk back to the umpire; then everybody present mixed it up in a free-for-all, the spectators adding to the fun by throwing stones. It was found that this sort of thing was reducing the strength of the army; so rules were made and the holding of tournaments was restricted to five localities in England. Kenilworth Castle was one of these, and its level, green tiltyard still exists.

The rules divided the hastilude into three types. The first and most important was the joust, which was singles—one against one. The joust almost always began with the familiar tilt or charge. From his own end of the field each man ran his horse at the other at full *tilt,* gripping the shaft of his spear under his right arm and seeking with it to unhorse his opponent, that is, knock him out of his saddle and over his horse's stern. To oppose this the backs of the saddles were made very high and there are records of knights being tied in place.

Ordinarily three of these "courses" were run. Then, if both men were still upright, they'd meet in the center of the field and exchange three blows with the mace or the battle-ax. Surviving this, and miraculously they often did, they'd exchange three sword slashes, either from the saddle or dismounted. If one champion was knocked

KNIGHTS JOUSTING IN A TOURNAMENT

46

out at any point there wasn't any doubt about the winner; otherwise, the judges decided the match on points.

The next hastilude in importance after the joust, and less technical but even more exciting, was the tourney. In this the knights drew up sides and fought a miniature battle, though still under rules; the chances for fouling were wonderful. The tourney created a terrific uproar and the spectators loved it.

The last of the three forms of the hastilude is something of a mystery. It was called the *behourd* and was some sort of exercise with spear and target, but exactly what kind isn't known. (In Maryland and Virginia something called a "tournament" is still held; in it the "knights" attempt to impale a small suspended ring on the point of a lance while riding at full gallop. It is interesting to wonder if this may be a descendant of the behourd.)

Everybody turned out for a tournament. There was a canopied gallery for the queen and her ladies and another for the king and his nobles. The commoners crowded the barriers around the lists and the boys unquestionably climbed trees. It was more than a mere "passage of arms"; it was a pageant, and as time went on it became less and less a combat and more and more a show, until it finally reached a pitch of silliness which the human race didn't exceed until it thought of flagpole sitting.

Since only knights could enter tournaments, the commoners staged their own games and combats. Individual bouts with the quarterstaff and with sword-and-buckler were common producers of cracked heads and minor scars. Tilting at the quintain was favored by those who could afford a horse or borrow one. To tilt at the quintain the contestant, riding at a gallop, aimed a blunt spear at a shield-shaped target which was hung at one end of a horizontal beam. The beam was pivoted on top of an upright post and at its other end hung a sandbag or a pail of water. The object was not so much to hit the target, which was large, as to avoid being slugged or drenched *after* hitting it.

At Easter the younger London set went in for tilting from boats on the Thames. This was the same game which is now played at summer camps with canoes, except that the boats were heavier and each was rowed by several oarsmen. A single combatant managed the padded lance in the bow, and the object was simply to push the opposing boat's lancer into the drink; sometimes they were a little rough about it. In the winter "jousts" were staged on ice, the charges being made on bone skates.

The quarterstaff was named from the way it was handled, not from its six-and-a-half-foot length which might well qualify it as a *whole* staff. It was straight and just under two inches thick for most of its length. The ends were a little

TILTING AT THE QUINTAIN

BOUT WITH QUARTERSTAVES

thicker and were usually loaded with iron.

In operation the quarterstaff was held in the middle in one hand while the other grasped it about a quarter of the way from one end, hence the name. The trick was to spin the staff this way and that, shifting the grip of the hands from quarter to quarter, thus delivering blows from unexpected angles and, at the same time, using the staff to ward off attack. It was quite a rugged sport. In *Ivanhoe,* the swineherd, Gurth, ended his bout with the miller by sliding his right hand from the middle of the staff down to his left at the quarter and delivering a haymaker.

The "Exercise of the Sword-and-Buckler" was a fencing match with slashing swords. The buckler was a small round shield of the type sometimes called a "target." It was about a foot in diameter and was provided with a handle on its back by which it was held in the left hand. The sword used was straight, tapered and double-edged, much like the knights' swords but shorter, about three feet overall.

For a couple of men thus equipped to stand slashing away at each other seems a strange amusement for a Sunday in the park, but that is

what it was. Spectators placed bets and championship bouts were held. Boys practised with sticks and homemade shields. The sport remained popular for several centuries.

Archery contests were frequent, and were usually held after mass on Sunday in every country churchyard, but we'll wait to a later time to examine them, when the record is less dimmed with age.

## Knights and Armor (1200—1300)

The knights of the early thirteenth century wore the same chain-mail hauberks that their fathers had worn at Hastings, except that the skirts were made longer, following the fashion of civilian clothes, and the sleeves were made full-length to protect the arms better. The sleeves ended in mittens of mail which had slit palms for getting the hand out. All armor now included chain-mail leg and foot coverings.

Each link of this mail was separately forged and welded. It would have been easier to make the links of wire but no one had yet discovered how to draw wire. A number of patterns of mail were in use, all of which are recorded in exact detail on tombstone statues and all of which are

THE "EXERCISE OF THE SWORD-AND-BUCKLER"

various combinations of interlocked rings. None of the actual material of these very early hauberks has survived. It rusted quickly and then, too, it must have been fine stuff for scouring pots. This tendency to rust caused the introduction of the sleeveless chemise or surcoat which was worn over armor to protect it from dampness. When chain mail became rusty, it was put into a barrel with small stones and coarse sand and rolled around the courtyard for an hour or so to clean it up.

Chief among the added protections however was the "heaume," or helm on the head. In its first form this was an iron pot in the shape of a flat-topped cylinder, open at the bottom, pierced or slotted for seeing and breathing, and weighing some thirteen or fourteen pounds. The helm was worn over the mail hood which was part of the hauberk, and often covered an additional iron skullcap called a basinet. The basinet was elaborated over the years into the headpiece we usually think of when we say helmet.

The weight of the first great helms was borne entirely by the head; later in this century, when helms had domed tops and hinged fronts, they were made deep enough to rest on the shoulders. A man fighting needed a lot of air, and since the slots in the helm were made small to keep spear points out, quite a few fighters smothered in their buckets. Smaller iron hats, some with brims, some with nose guards, were used at times in real battles without the helm and were called helmets, which means "little helms."

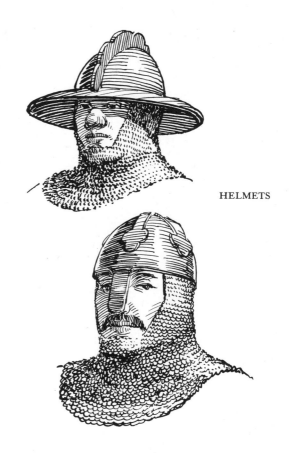

HELMETS

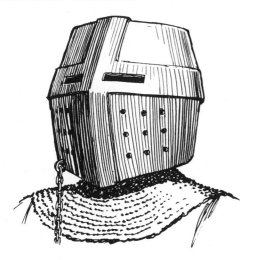

IRON HEAUME

The knights' hand weapons had changed little since Hastings. The lance was somewhat longer, perhaps a bit heavier, but it was still a simple pole. The sword was the same except that it now had its cross hilt turned down a little, and it was hung on a fancy draped belt which put the hilt right over the knight's breadbasket. The misericord began to be worn on the right side. This was a dagger which has been called the "dagger of mercy" 'with the idea that it was used for the quick dispatch of a suffering loser; actually it was the persuader which, presented point first to a fallen foe, impelled him to plead for mercy and come across with a healthy ransom.

In war, though not in tournaments, the falchion became deservedly popular. It was a real snicker-snee—a sword, but built more like a knife or a cleaver, and nearly three feet long, with a single, curved cutting edge supported by a very thick-backed blade, which gave it weight and authority.

49

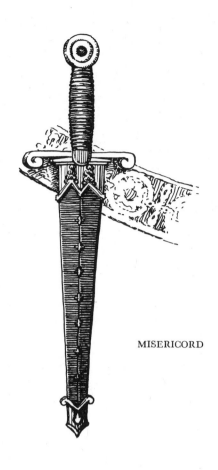

MISERICORD

FALCHION

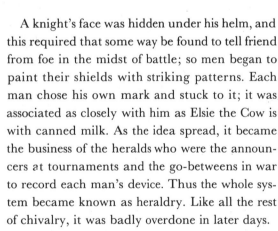

A knight's face was hidden under his helm, and this required that some way be found to tell friend from foe in the midst of battle; so men began to paint their shields with striking patterns. Each man chose his own mark and stuck to it; it was associated as closely with him as Elsie the Cow is with canned milk. As the idea spread, it became the business of the heralds who were the announcers at tournaments and the go-betweens in war to record each man's device. Thus the whole system became known as heraldry. Like all the rest of chivalry, it was badly overdone in later days.

In the illustration, the knight is in complete chain mail with a basinet under his hood and added plates at his knees. He wears his sword over his long surcoat. His helm is under his right hand and with his left he is about to place the *guige* (strap) of his shield over his head. A squire is holding his horse while pages help with his equipment.

This is a wealthy knight of some importance, perhaps a baron. His horse is *trapped* with expensive chain mail, and the spear his squire is holding bears not a mere pennon, but a banner of the knight's heraldic arms.

Chain mail was heavy and the plate armor of the fourteenth century was much heavier, so the horses ridden by armed knights had to be big and strong. Fairly nimble Spanish horses were favored in the thirteenth century, but later, in France, Flanders and England, special horses were bred which could carry more weight. Their descendants, the Percherons, Belgians and Clydesdales, are the best draft horses in the world today.

Nearly everybody in the Middle Ages belonged to some trade guild, and the brotherhoods of chivalry have been called "The Guilds of the Horse Butchers." It's true that they went for each other's horses. A knight's horse was his fighting platform, as well as his most vulnerable point. Kill or cripple his horse and you had your knight where you wanted him. One device for accomplishing this was to sow the field where an enemy would charge with little four-pointed metal gadgets called *caltrops*. The points of a caltrop were so arranged that one of them always stood straight up.

In order to protect their horses from spears and arrows the wealthier knights began to drape them with a "trapper" of chain mail. Hoods with eyeholes covered head and neck and hung nearly to the ground on both sides; a blanket of mail equally long was spread over the horse's rump. Since this rusted, it came to be covered with a drapery on which the knight often painted his heraldic device. Poorer men protected the horse with quilted cloth.

When the Crusaders took their strong horses to Palestine, they suddenly discovered the advantages of military mobility. The light-armed, swift-mounted Saracens rode circles around the plow horses and hit them from all sides. All of the Saracen weapons were planned for these hit-and-run tactics and hence were very different from the European ones. There is a famous (and probably false) story which illustrates one of the differences. At a truce meeting Richard I showed off his bull strength by severing an iron bar with one sword stroke. The Sultan Saladin then set everybody's teeth on edge by slicing a sofa pillow in half with his curved, razor-sharp scimitar. However, a scimitar was not too good for slicing an iron hat, which

CALTROPS

A THIRTEENTH-CENTURY KNIGHT ARMING FOR A TOURNAMENT

may be the reason the Crusaders were able to maintain a toehold in the Holy Land.

## Medieval Armies and "Gyns" (1300—1400)

The medieval years were rough times. Just as in the twentieth century, citizens were often attacked and robbed, so nobody went about in public without arms of some kind. The tradesman, priest or knight in civvies wore a short sword (or long dagger) called a *baselard* suspended from the belt of his gown. Lesser men carried the quarter staff and some sort of dirk. Gentlemen's daggers were often hung around their necks or carried in a pouch. Even women wore fancy little daggers fastened to their girdles.

Everybody was in some way involved when an army had to be raised, but to judge from the records, some of the service was on the casual side and the draft boards apparently weren't very tough. At least one archer who went to the wars was required to stay with the king's army "until he had shotte away hys arrowes," and another took with him a flitch of bacon and was a soldier only until he had finished eating it. However, the British yeoman, because he was allowed some self-respect, was gradually becoming a foot soldier who made his weight felt. On the Continent where the peasants had *no* rights and were treated like cattle, they behaved accordingly.

Medieval armies didn't go in heavily for organization but there was some attempt at subdividing the mob. An army was made up of three *battles,* each consisting of a greater or lesser number of *routes* or *retinues* containing from twenty-five to eighty men each. A route might consist of the retainers of only a single lord or knight-banneret, or several such retinues might be grouped together as a route. Command of an army so organized was badly complicated by the fact that each soldier would obey only his own landlord, each landlord felt responsible only to the baron from whom he held his fief, and this nobleman in turn disregarded everybody but the earl to whom he had sworn fealty.

In battle, archers and crossbowmen were usually placed in front of the mounted knights, and after the footmen had shot their arrows, the cavalry would charge through them; or in France, sometimes over them—after all, they were only dogs of peasants. At the battle of Crécy the Genoese crossbowmen, when they had shot their bolts *for* the French, were ridden down *by* the French and suffered more hurt from the knights of their own side than they did from the English.

Up to that battle there had not been, since the heyday of Rome, any question as to the superiority of mounted knights over foot soldiers, but at Crécy the English longbowmen, standing on their own two feet, began to embarrass the horsemen. They caught the French knights in a crossfire and annihilated their horses with carefully placed arrows. On the same day, the Black Prince made some of his knights remove their spurs and fight on foot with shortened lances. He seems also to have used a cannon, but we'll talk about that later.

The Greeks and Romans planned whole campaigns in advance but strategy was a forgotten idea in the Middle Ages. Attacking and defending armies seldom knew where they were or where the enemy was; sometimes they played hide-and-

FOOT SOLDIER

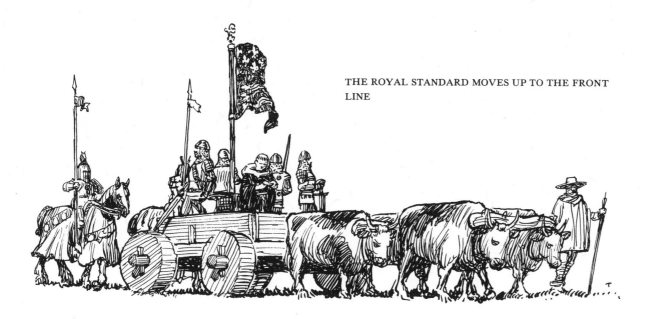

THE ROYAL STANDARD MOVES UP TO THE FRONT LINE

seek for weeks and went home without ever finding each other. There were no maps. Often an invading army arrived in front of a town either with no idea what town it was, or else quite sure that it was some other place entirely.

Crécy was fought in 1346, and in the battle the English used some tactics planned in advance. Up to then nobody had thought much about such things. Most battles were a series of personal combats between knights, and the nearest they came to concerted action was to help a friend out of a tight spot now and then. Because of the resulting confusion it was necessary to have rallying points of some kind. Banners marked with coats of arms or "badges" were one device for this. Similarly the king had his standard and a picked group to defend it; often this banner was flown from a staff mounted on an ox-drawn wagon. As long as the standard could be glimpsed above the fight, the men who followed it knew they were not completely licked.

Men also needed to be seen above the melee and recognized, so knights began to wear crests on their heaumes. These were carved from wood or molded of leather, and usually represented some animal or object which served as a secondary trade-mark for the knight. The first crests were just little fans of feathers, but when it was realized that they made their wearers look taller and more imposing, the crests became very elaborate and very, very high.

Battle cries were popular, and in addition to whooping up the fight, like the Japanese "Banzai," they helped to rally men around a leader. It's difficult to imagine a battalion of American GI's attacking with shouts of "Liberty and the United States" or "Ha, Eisenhower!" At times they shout as they attack but the words are quite different.

The *trompe,* the *oliphant,* the *claironceau* and a dozen other horns with handsome names were sounded for rallying, and the *tabour* or *tambour* (drum to you) was beaten for the same purpose

CRESTED HEAUME

53

though it's hard to see how it was heard over the racket which prevailed.

Siege was no small part of medieval military operations, and the reduction of a strong castle was a long, weary job. It took Oliver Cromwell's Puritans three years to capture Corfe Castle, even though they had cannon and muskets by then and the garrison of the place consisted of little more than the doughty Lady Bankes and her serving wenches.

The oldest and simplest of siege weapons was the battering ram, but against a ten-foot-thick stone wall it wasn't worth a hoot. Medieval armies used them against town gates when they could be worked. The "mouse" was a drill as the Roman terebra had been. It was rotated by simple handles and it, too, had small chance against a stone castle.

The fighting tower, officially called a *beffroi,* was known to the soldiers of the Middle Ages as a "cat." Usually it couldn't be brought close enough to a castle's walls to be effective, since it was necessary to fill in the moat to make a passage for it. Under the name of *escalades,* ladders of various kinds were used to try to climb over high walls; but a man on the top of a tall ladder has no perceptible advantage over a group of armed men above him who are trying to push his ladder over and to stick him full of spears and arrows.

In view of these things it isn't hard to see why the besiegers of castles depended mainly on missile-throwing *gyns* for their assault. In the old days of wooden castles the *springal* did very well. This was a version of the Roman *falarica* which propelled darts by spanking them with a springy timber. Usually the darts were incendiary and kept the defenders busy putting out fires. Lead came to be used as a material for roofing castles largely as a protection against attacks of this kind. Because it was the lightest of projectile-throwers and because ships were highly inflammable, the springal was for centuries an important naval weapon.

Quite a bit more powerful than the springal was the *ballista* which was sometimes called an *arbalest,* but because the latter name is also applied to the high-powered crossbow, we'll stick to the older term. This gyn was really like a catapult and was never as large or as powerful as the Roman ballista had been. The medieval one had no skein of twisted fibers; it was simply an enormous bow, bent by a windlass. The Roman sliding trough

SCALING LADDER

SPRINGAL

54

BALLISTA

was missing; the trigger was a simple forked slip-hook. The medieval ballista pitched a javelin in very good style however and was probably the most accurate of the siege weapons.

The *mangonel* was mounted on four wheels and was familiarly known as a "nag" for the same reason that the Romans called it a "wild ass," because it kicked up its rear-end when it was fired. The mangonel threw stones. In later years the word was shortened to "gonne," and since the first cannons also threw stones, they became known as gonnes too. In principle the mangonel was exactly like a Roman onager; but in practice it was a much cruder machine and puny in comparison. Its great advantage was that it could be moved; it was a "field gun." The skein which was its "propellant" was never as efficient as the Roman version. It seems usually to have been equipped with a scoop for holding its projectile and was seldom given a sling; on the other hand the *trebuchet* almost invariably had a sling.

The trebuchet was the heavy howitzer of medieval artillery. Usually it was so large that it had to be constructed at the scene of action; its *verge* or beam ordinarily was made of an entire tree. While it never equaled the range of the classic ballista

and seldom did walls any real damage, it turned in a performance good enough to scare the daylights out of the residents of a castle.

To understand the trebuchet, picture a seesaw with one short end and one long one. Put three-hundred-pound Uncle Henry on the short end and hold the long end down while you put thirty-pound Junior on it. Now let go and watch Junior sail over the house—never mind what happens to Uncle Henry.

On the actual gyn the seesaw was pivoted at its balance point, and instead of Uncle Henry there **was** a large box of stones and earth hinged to a fork

MANGONEL OR "NAG"

LARGE TREBUCHET

on the short end. The means of setting and releasing were like those of a mangonel except that the windlass was under the main trestle where the skein-winch would be on a mangonel, and the setting ropes were led to it under a roller at the back of the base frame. The trigger was often a large hook which held the verge itself, making it necessary to unhook the setting ropes before shooting. The stone projectile was placed in its sling on a long platform which began just back of the windlass. It was snatched from here by the released verge, and swung in an arc at the end of the pole. At a point about two-thirds of the way up, one side of the sling slipped off the end of the verge and the stone sailed free, following a high curve to its target.

In large trebuchets there was nothing to stop the verge after the projectile left it. The weight swung between the trestle legs; the arm threshed forward

56

and back and came to rest standing straight up. Apparently it was then necessary to shinny up the verge and hook the ropes onto the pole again after every shot! The weight on a small trebuchet usually dropped to a rest and lay there, the arm never reaching an angle much higher than sixty degrees. That was all right because with a sling, the stone had already left when the arm got that far up.

Trebuchet is the French name of this gyn; used of course by the Norman aristocracy, it has somehow remained in modern English. In the English of its own day it was called a "trip-gate" or a "trap-gate." A stone from it was "trapped" at the target and we still practice *trap*shooting at small projectiles thrown by a machine.

The trap-gate was used as a siege weapon well into the sixteenth century without any change of principle. Metal bearings improved its operation, and the substitution of a metal weight for the old box of junk made it handier and more permanent. However, it's doubtful if the ornament lavished on it added even an inch to its range or accuracy.

You will have gathered that siege engines had small effect on strong castles beyond annoying the tenants; sometimes the besiegers, knowing this, would concentrate on being as annoying as possible. Along with their rocks and spears, they would toss in the carcasses of very dead animals or, if they had caught one, a live prisoner, sometimes the courier sent by the castle to bring help. People weren't squeamish about little jokes of that sort in "the good old days."

The besieged castle had gyns of its own which operated from the court. They'd have been more effective from a higher point but the towers wouldn't stand the shocks of discharge. In their day the Romans had built special towers which were solid all the way up, expressly for mounting ballistas, but in the Middle Ages nobody remembered that or thought of it.

We've mentioned mining several times as an effective way of toppling walls by digging under their foundations. The work was done in the way a coal mine is dug today. As the tunnel advanced, it was shored up with heavy timbers. When the digging had progressed far enough to be directly under the masonry, the castle walls were supported on the shoring; then the diggers soaked their timbers with oil and pitch, set fire to them and went back to camp. Unless the defenders could find a way to flood the tunnel, the wall fell down when the shoring burned away.

Attempts also were made to set wooden gates afire with oil and pitch; so slots were angled in the castle wall above the gates in such a way that water poured through them would drench the outer face of the wood.

## Longbows and Crossbows (1300—1400)

In the illustration on page 58, the nearer man is about to release a long flight arrow at the enemy's rear rank; the other man is starting a livery arrow point-blank at the opposing crossbowmen. Actually, archers shot from *behind* the sharpened stakes of their anti-knight defense rather than from among them; but showing them this way made it simpler to get both into the same drawing. Those handles projecting from both men's backs belong to the mauls to which archers resorted when their arrows were gone.

Bows weren't new in England at the time of the conquest but the Normans made greater use of them in battle than the Saxons did. The Norman bow doesn't seem to have been the "great bow" or longbow which gave England her undisputed mastery in archery, but a shorter weapon which, to judge from the way it is shown drawn to the chest in old pictures, couldn't have approached the longbow in range and power. Yet six-foot bows were not new. The Egyptians had them, and they've also been found in the hulls of old Norse boats which were buried not later than the year 800.

Just when the English bow was lengthened and a better method of shooting was developed isn't known. When the longbows of some outlaws made his peace officers look foolish, Edward III had the idea of encouraging archery practice and forming companies of bowmen for his army; but this was around 1330, only a few years before the battle of Crécy, where the longbow startled the world. A lot of bow shooting must certainly have gone on for a long time before that.

The ordinary longbow, or "livery bow" as it was sometimes called, was supposed to be as tall as a man, but some of them were six feet four inches long and one monster, called *Little John's Bow*, still exists which is six feet seven! The livery bow "weighed" about a hundred pounds—that is, it took that much pull to draw it. Most modern bows weigh about fifty or sixty pounds, if they are heavy. The old longbow was a self-bow made from a *stave* of yew, basil, wych-elm, ash or hazel. Of these yew was by far the best, but the old phrase, "a bow of good English yew," is misleading. Much better yew than the native kind was brought to England from Italy and Spain by the Venetians, and it was imported from very early times.

A longbow was about an inch and a half wide and an inch and a quarter thick in its middle,

where the hand grasped it. The back, away from the archer, was flat, and the belly, facing him, was nearly half-round. Both ends tapered evenly and were capped with a bit of horn in which a notch was cut to hold the bowstring. The careful selection and shaping of the wood had much to do with the merits of a bow as a weapon, and the expert craftsmen who did the work were called *bowyers*.

The English archer used a bowstring (sometimes he called it an arrow-string) of hemp carefully whipped with light linen cord. Against it he set the nock (end notch) of an arrow with a light aspen-wood "stele" or shaft; a metal "pile" or head; and "fletched" or "flighted" with the halves of three goose feathers near the nock end. This was the "cloth yard shaft." A cloth yard was thirty-seven inches; only a very tall, very strong man could draw the bow which took an arrow of that length. The arrow was generally assumed to be the length of a man's arm, or half the length of a bow; but flight arrows for distance are known to have been longer than the ordinary sheaf arrow.

To shoot, an archer held his bow at the full length of his left arm, standing with his feet a little apart, his heels in line with the target. Nowadays the heels are nine inches apart but in the Middle Ages they seem to have straddled a little more. The bowman's body wasn't turned at all but faced as his feet did; his head was turned sharply to the left, facing his mark and sighting it over the pile of the arrow. The nock end of the arrow lay against the middle of the string (the place was marked) and was held there lightly between the

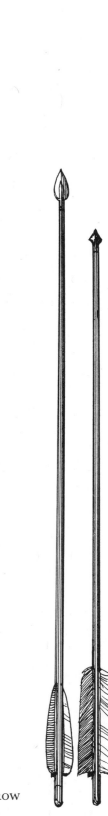

THE TWO-FINGERED DRAW

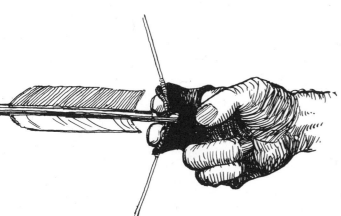

LONGBOW

*Unstrung*      *Strung*

FLIGHT ARROW AND LIVERY OR SHEAF ARROW

right forefinger and the middle finger which lay across the string protected by a glove or a leather tab. After 1500 the third finger also held the string. Either way was more effectual than the ancient pinch draw. Arrow and string were drawn back together until the nock lay directly under the archer's right eye, just at the angle of the jawbone. To draw a thirty-seven-inch arrow to its full length, most men have to draw it *past* the jawbone and it's hard to sight it accurately. The pile end of the arrow lay to the bow's left, resting on the bowman's left hand. In an almost continuous motion the archer drew, aimed and loosed; the bow did the rest.

with increasing distance, passes through a point where it coincides with the bull's-eye, and for long range is actually above the target. It is the selection of his aiming points plus a number of other things that makes a good archer.

An arrowsmith made heads only. The arrows themselves were made by a fletcher. Arrowheads varied widely in shape but there were two main divisions: the pile and the broad. The pile was sometimes leaf-shaped and sometimes lozenge-shaped, but more often it had a quite blunt point and was little larger in diameter than the shaft of its arrow. This was the war head which could pierce chain mail or kill a horse at two hundred

DIAGRAM: AIMING ARROWS

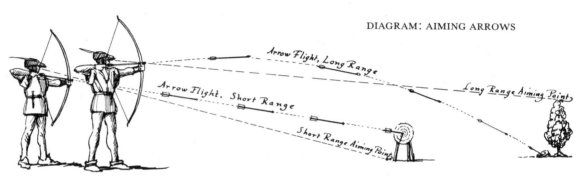

An archer can't sight along his arrow as one sights along a gun barrel, directly at the target. It isn't physically practical to draw a bow with the arrow on the eye level. There's only one right height to use a man's strength effectively and that at the level of the angle of the jaw. The sighting eye is about four inches above that. It isn't too hard to learn the feel of lining the shaft up horizontally by drawing to a point directly under the right eye, but in the vertical plane the eye looks downward at the arrow. If impulse is followed and the point is raised until it appears to center on the bull's-eye, the shot will pass high over the target at any ordinary range. So, for distances up to sixty yards, archers learn to pick an "aiming point" below the target.

Sixty yards is about the distance an arrow will fly straight without dropping; beyond that allowance must be made for the effect of gravity by shooting at a higher angle. The aiming point rises

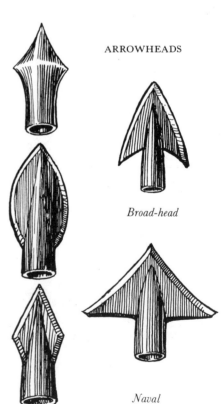

ARROWHEADS

*Broad-head*

*Naval*

*Pile*

yards; at closer range it would puncture ordinary plate armor. The standard broad head was quite sharp and had two wide barbs; it was much favored for hunting. A special *very* broad head was used in sea fights to cut sails and rigging. There were other trick shapes, such as the fork head and a crescent shape, which were fancied for special purposes.

The longbow has provided remarkable demonstrations of the penetrating force of an arrow. Men have been pinned to their mounts by a single shaft through both legs and the horse! At close range arrows from longbows have been shot through oak doors three and a half inches thick, and an arrow from a bow two hundred and twenty yards away has been driven entirely through a one-inch oak plank.

Every English king from Edward III to Henry VIII encouraged archery and gave prizes for it. At some periods archery practice was compulsory, and the minimum distances from which a man might shoot at a target were fixed by law. In France, after the great successes of the British bowmen, there was a movement to encourage archery as a sport, but it was soon quashed because the nobles were afraid of so dangerous a weapon in the hands of the peasants. The English yeoman, who was a free man and who was on reasonably good terms with his overlord, could usually be trusted not to put a gray-goose shaft into the boss's back.

Target competition was keen. There were earthen butts in every hamlet and rounds were shot on Sundays and feast days. For variety the popinjay was put high on a tower and used as a target. The popinjay was a brightly painted wooden bird, supposed to look like a parrot. Clout shooting, in which the target lay on the ground and was shot at from a long distance, was popular. In this game markers stayed near the target behind shields and came out to signal the success of shots with flags. To indicate a shaft "in the clout" the marker fell flat on his back!

Another archers' amusement was the curious game of "rovers," which has been likened to golf because the players moved across the fields, shooting from one target to another. Special fields were set aside for the sport. Each target had a name

and was a permanent fixture set up at the expense of some important person to cull favor with the public.

Robin Hood certainly did *not* shoot an arrow a mile; but many of the archery stories which have come down to us we can take for gospel because they have been approached, duplicated or beaten in our own day. Only a few years ago a Dr. Pope was able to shoot seven arrows upward, loosing the seventh before the first hit the ground. Hiawatha shot ten, but we have only Longfellow's word for it. The present flight shot record exceeds any authenticated distance in history. Locksley's one-inch peeled wand in *Ivanhoe* has been split many times by young men who also drive automobiles. As late as 1793 the longbow beat the musket for accuracy; and in 1924 General Thord-Gray made twelve pistol experts look silly by putting seventy arrows out of seventy-two into a twenty-six-inch target at eighty yards; the pistol men all shot nearer and scored worse.

But let's get back to the fourteenth century. The military longbowman was a stout fellow, selected for his size and strength. Normally he wore no armor except an iron cap and a quilted tunic; a few lucky ones had mail shirts or boiled-leather chest pieces. The archer's hair was cropped short to keep it clear of the bowstring, and if he had a beard, he held it in his mouth while shooting for the same reason. His left wrist was protected from the snap of the bowstring by a leather wristlet called a bracer, and he held a leather tab in his right hand to keep the string from cutting his fingers. For arms aside from his bow, he sometimes wore a sword, but almost always he carried on

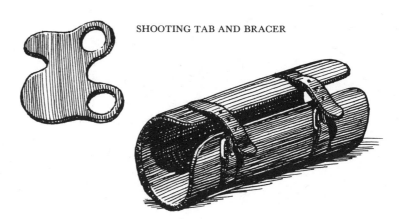

SHOOTING TAB AND BRACER

his back a twenty-five-pound maul with a four-foot handle and an iron-bound head of lead.

On the battlefield the archer often set up a stake sharpened at both ends and leaned it forward at an angle, to protect himself from cavalry charges. Or he might have with him a soldier who held an over-size shield called a *pavise* to cover the archer while he shot. At sieges the bowman was more likely to work behind a *mantlet* which was a rectangular wooden screen with a prop hinged to its back.

Perhaps the outstanding point about these men, as about the American riflemen of later times, was

MANTLET

PAVISE

that they picked their targets and hit them; they didn't just cut loose in the general direction of the enemy. Any qualified archer was expected to shoot a dozen arrows in one minute at a man-size target two hundred and forty yards away—and *hit* it with all twelve.

At his waist the archer carried a sheaf of two dozen arrows, "four-and-twenty Scotchmen in my belt" was his way of putting it. Eighteen of these would be sheaf arrows, the other half-dozen were flight arrows. In action the whole bundle was shaken out and the arrows lay on the ground points outward, near, some say under, the archer's foot. Some very early drawings show a quiver for holding the arrows behind the right hip, but its use was abandoned. When a man had shot all his arrows, he sometimes could advance and recover them or others from the bodies of the slain, then shoot with them a second time at the rear guard.

The English archer three times in a row, at Crécy, at Poitiers and at Agincourt, decimated "the flower of French chivalry," and so doing he put the foot soldier back on the military map for the first time since the decay of the Roman legions.

There was also the crossbow. It is said to have been used at Hastings but there is no representation of one in the Bayeux Tapestry. The crossbow involved no new principle. It was a very small, very stiff bow, set crosswise at the end of a staff or stock. In effect a small ballista. Its great advantage was that it could be drawn ahead of time and could hold its draw while the bow was aimed, and it could be raised to eye level and sighted. Actually, the early ones were pretty poor weapons, but the Pope considered them too murderous for "Christian warfare" and pronounced an interdict against them in 1139. The use of them against infidels was permitted, however. Richard the Lion-Hearted disobeyed the edict, and people generally felt that it served him right when he was killed by a crossbow bolt.

There was a metal stirrup on the front of the crossbow stock. When he wished to set the bow, the archer put one foot into the stirrup, then grasped the bowstring with his hands and strained it back far enough to hook it over a little catch called the "nut." The nut was something like a spool with a notch in it. The notch held the string until pulling the trigger allowed the spool to rotate enough to let the string slip off. This device was never improved upon for its purpose.

The upper drawing shows the nut and trigger of an early crossbow in a "set" position with string on nut and bolt against string, all ready to be "let off." The ivory nut is kept in its metal socket by a catgut lashing; later a pin was used for this.

The lower drawing shows the nut and trigger still "set" in the same position, but the wood of the stock has been cut away to show how the spring holds the end of the trigger bar in a notch on the under side of the nut to prevent the nut from turning until the trigger is squeezed. A little wedge of metal set in the nut takes the wear of the trigger.

Mostly the crossbow discharged short arrows called bolts. Later, because they had square heads, these were called quarrels. The heads were iron. The short, thick shafts were wood and the "feathers" were leather or paper. Bolts for hunting were nicely finished and had three vanes like a standard arrow, but two of them were directly opposed so the bolt could lie flat in the groove of the stock. War quarrels were quite roughly made and had only two vanes, which were made as one and inserted into a saw-cut which was lashed tight behind them.

It isn't known just when the first gadget appeared for helping to set a crossbow. It was called a *belt claw* and that's what it was: a double hook hung from a belt. By hooking it on the string and sticking a foot in the stirrup, a man could take

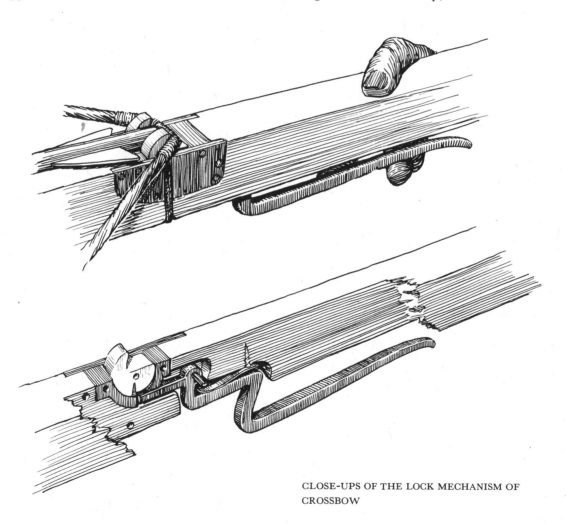

CLOSE-UPS OF THE LOCK MECHANISM OF CROSSBOW

63

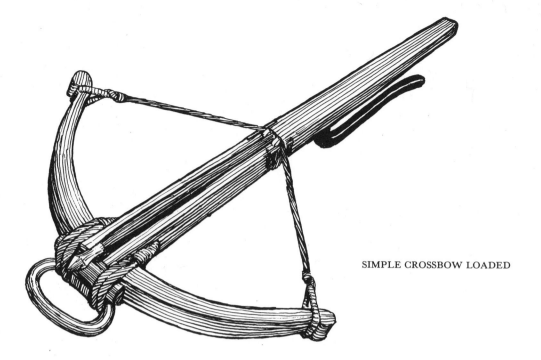
SIMPLE CROSSBOW LOADED

ARCHER DRAWING CROSSBOW

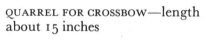
QUARREL FOR CROSSBOW—length about 15 inches

BELT CLAW FOR DRAWING CROSSBOW

OPERATING THE CORD AND PULLEY

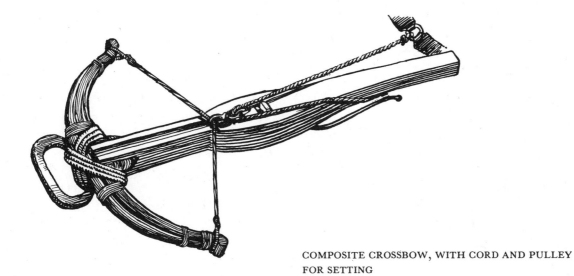

COMPOSITE CROSSBOW, WITH CORD AND PULLEY
FOR SETTING

advantage of his strong leg muscles and so set a bow too stiff for his arms.

After the bow was set the bolt was placed in the groove with its square end (a bolt seldom had a nock) lying between the two lugs of the nut and against the string. To shoot, the bowman raised the stock to eye level, sighted directly at his target over the knuckle of his right thumb and squeezed the trigger.

There were disadvantages. The crossbow was heavy. It was slow. It could shoot only one quarrel while the longbow was delivering six arrows. Though it did not require a specialist and was more accurate at short range in the hands of the average soldier than the longbow was, yet its range was *so* short that it often couldn't be brought close enough to the enemy to bother him. The bowstring was ordinarily twisted of sinew or gut, and in damp weather it became entirely limp and useless; armies which depended heavily on the crossbow found this to be something worse than a mere nuisance.

The constant efforts which were made to increase the power of the crossbow presently resulted in adding whalebone and animal tendons to the basic yew frame. This composite was an improvement, and it was now too stiff for a man to set even with a belt claw. A simple little purchase was invented to help with the job. This was a short piece of rope running through a pulley which had a hook on it to engage the bowstring. One end of the rope was attached to the archer's belt; the other end could be hitched to the stock of the crossbow.

To set his bow the bowman with his foot in the stirrup, bent forward, hooked on to the string and the stock simply straightened up. In addition to taking advantage of the strong muscles of the lower back and hips, this gave a mechanical advantage of nearly two to one; so a weak soldier could set a strong man's bow and crossbowmen didn't have to be exceptional physical specimens.

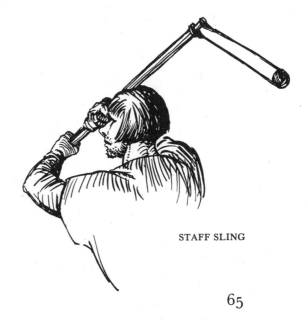

STAFF SLING

65

## Knights and Guns
### (1300—1400)

When good archery came into the fight, the mounted man-at-arms discovered that he needed something more than chain mail to save his skin and his reputation. Some of his soft spots were already protected by iron plates, but now he added more of them over his mail, until by the end of the century he was entirely encased in iron. Then he discarded the mail.

BASINET WITH MOVABLE VISOR. The pin had a spring behind it and held the visor open

The great helm or heaume proved to be just the right thing for jousting; in fact, it had been made still bigger, until its weight rested entirely on the warrior's shoulders, but in a real fight it was too cumbersome and could be knocked off too easily. There had to be head protection in battle, so the basinet was gradually improved from a simple skullcap into an elaborate helmet with a movable facepiece or visor. This was closed only in actual combat and had breathing holes in its right side but none in its left, so that a lance point could slide off without catching in them. The earlier basinets had a kind of skirt of chain mail laced to their

lower edges to protect the knight's neck, but it didn't work too well and was replaced by an iron neckpiece which could be fastened to the body armor.

When two knights charged each other, whether in the lists or in battle, the one whose lance point outreached his opponent's had the advantage, like a prize fighter with long arms. So everybody made his lance a little longer and then longer still, until the thing was stopped at the maximum length that a man could handle on horseback, which proved to be about fourteen feet. A little round hand shield was added to the lance and the butt end was thickened just back of the handle to improve the balance. Even so, lances were so clumsy that little cleats began to be attached to breastplates to help in holding them steady. Shields, too, often had a notch in the upper right-hand corner, in which the lance could be rested. Swords changed little except to get longer.

The advances of archery and the efforts of the bowmen to get at the knight through his mount led to the adoption of heavy horse armor. The horse's head, neck, chest and rump were covered either with chain mail or with iron plates having an ornamented cloth over the metal to discourage rust. Nothing less strong than a draft horse could any longer serve a knight.

All this horse armor, together with the man's iron fighting suit, was not only heavy but very expensive. The law forbade the export of armor and controlled its price, but it was still so valuable that a knight usually made a special bequest of his armor in his will, and it customarily descended from father to son.

The tournaments took on a dizzy romantic tinge, with bits of female clothing tied to helmets and the winning champion's lady crowned as the Queen of Love and Beauty. The Masked Marvel act was a favorite. A strong knight would appear disguised (an end accomplished simply by giving his shield a coat of paint), and would challenge all comers. Sometimes a group would get themselves up as "Knights of the Round Table," each taking the name of a knight in the story, and *they* would take on all comers. This act was carried so far that they dined together at an actual Round Table with each man's place lettered with his play-acting

66

THE BLACK PRINCE IN ARMOR halfway between chain mail and plate

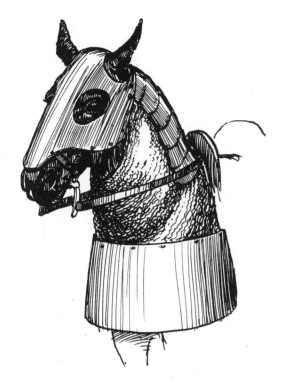

ARMORED OR "BARDED" HORSE

name. One of these table tops still hangs on the wall of Winchester Castle Hall, or did twenty-five years ago.

Since it was sincerely believed by all that God invariably would be on the side of justice and truth, lawsuits both civil and criminal were customarily settled by combat. Many times the actual fighting was done by champions hired by the principals in the case or appointed by the court. In the case of a prosecution by the State, there would be a King's Champion against the offender or his substitute. Few doubted that justice was accomplished, and they often ended the matter by hanging the loser if he survived the fight.

About the middle of the fourteenth century a type of organization appeared which scourged Europe for nearly three hundred years. This was the Free Company. Its members weren't knights but they were frequently armed like knights. A Free Company was really a private army, owned and equipped by its leader and rented out to the highest bidder. It was without allegiance or conscience; a higher offer might make it change sides overnight. If there was *no* offer, the Free Companions simply lived off the country, taking what they needed and anything else that was loose at one end. France and Italy were devastated by these criminals; some called themselves English or Scottish and a few actually were.

These gangs and one other thing eventually ended chivalry and feudalism. The other thing was gunpowder. There's little doubt that it was discovered by the Chinese long ago, but they used it only for firecrackers. The idea of propelling a projectile with it has been credited to the Arabs and to a German monk named Schwarz who blew himself up finding out about it.

Perhaps the very first cannon was the Arabian *madfaa* which was a deep wooden bowl holding powder; the cannon ball didn't enter the barrel at all but was balanced on the muzzle and popped off by the explosion. The *pot de fer* was better. It was an iron bottle with a narrow neck. The powder filled the bottle itself and an iron arrow, wrapped with leather for a tight fit, was rammed into the neck. Near the bottom of the bottle there was a little touchhole through which a red-hot wire was thrust to set off the explosion.

From the *pot de fer* the first true cannon were developed. They were simply pipes, closed at one end and firing stone or lead balls an inch or so in diameter. Some guns may have been made of wood,

POT-DE-FER, THE FIRST METAL CANNON

67

bound with iron. There was no carriage or framework of any kind to support them; they were simply laid on the ground with a heap of earth under the muzzles to aim them up in the air a little. Sometimes they fired crossbow bolts.

Though the French knights were outraged on principle when the English used cannon at Crécy, actually a man in armor was quite safe in front of these early ones. Their noise was impressive but the missiles they threw had little more punch than a man's arm could have given them; they bounced off plate armor and hardly dented it.

It was soon realized that this popgun wasn't much of a weapon. There was no immediate way to give more power to a small projectile, so it seemed best to make the whole business larger. There was no way to cast a large iron ball, so

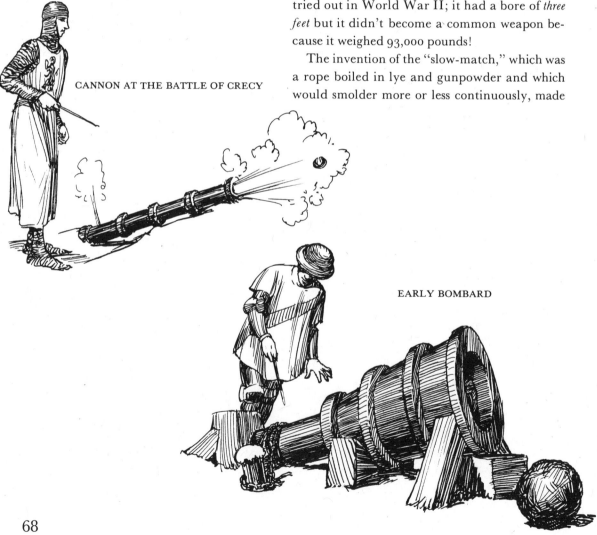

CANNON AT THE BATTLE OF CRECY

EARLY BOMBARD

each ball was cut laboriously from stone. It was not a case of slow growth from little to big. The little ones were flops, so enormous ones were built at once.

These big fellows were called bombards. The first ones were short-barreled and much smaller at the breech end than at the muzzle. With a bore that shape, the exact size of the stone ball was not important; the smaller it was, the further down the barrel it went.

Since the short bombards hadn't nearly as much range and force as a trebuchet, longer and larger barrels were tried. These were built up of parallel iron bars bound with hoops, on exactly the same principle that a beer barrel is made; and the bores of the larger ones were considerably greater than those of any guns commonly in use today. The balls were twenty or even twenty-five inches in diameter. A mortar christened "Little David" was tried out in World War II; it had a bore of *three feet* but it didn't become a common weapon because it weighed 93,000 pounds!

The invention of the "slow-match," which was a rope boiled in lye and gunpowder and which would smolder more or less continuously, made

68

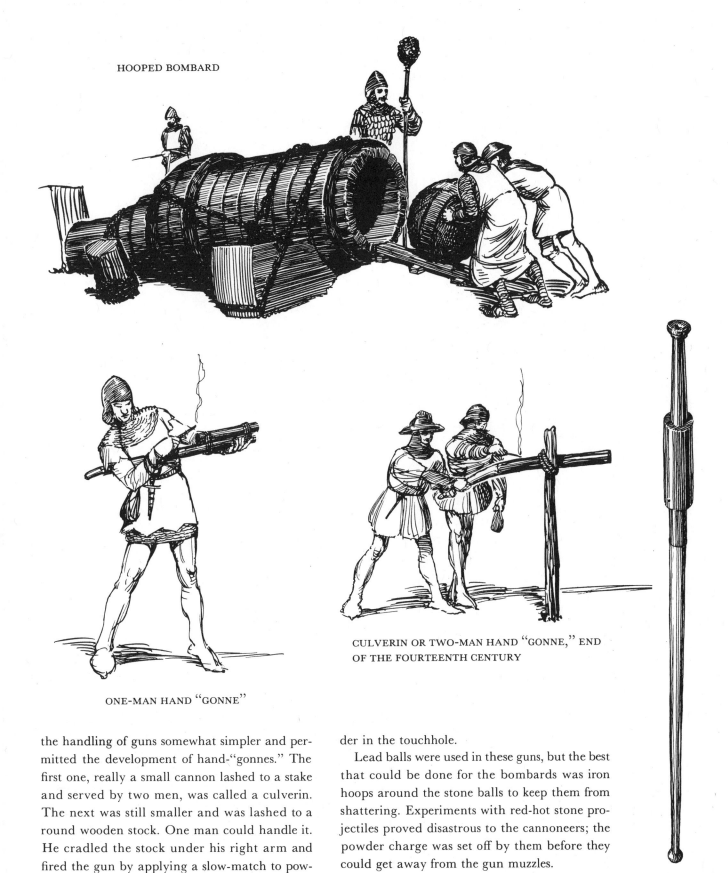

HOOPED BOMBARD

ONE-MAN HAND "GONNE"

CULVERIN OR TWO-MAN HAND "GONNE," END
OF THE FOURTEENTH CENTURY

the handling of guns somewhat simpler and per-mitted the development of hand-"gonnes." The first one, really a small cannon lashed to a stake and served by two men, was called a culverin. The next was still smaller and was lashed to a round wooden stock. One man could handle it. He cradled the stock under his right arm and fired the gun by applying a slow-match to pow-der in the touchhole.

Lead balls were used in these guns, but the best that could be done for the bombards was iron hoops around the stone balls to keep them from shattering. Experiments with red-hot stone pro-jectiles proved disastrous to the cannoneers; the powder charge was set off by them before they could get away from the gun muzzles.

HAND "GONNE," END OF THE FOURTEENTH
CENTURY

## Proof Armor, Arbalests and Breechloaders (1400—1500)

This seems to have been a century, not of innovations but of refinements and improvements. Armor, crossbows, guns, all were improved. The complete plate armor of the knights grew to be very fancy but it was also very good. The thrust of the cloth-yard arrow and the punch of the crossbow bolt demanded good metal to resist them. Presently the armor became so heavy that a man once down was helpless.

The joints of armor came to be made very tight to resist rapier thrusts. There's a story of some knights whose enemies came upon them unhorsed, supine and helpless and undertook to finish them off with daggers. Their points could find no crevice in the tightly fitted armor through which to reach a vital spot; so an ax was borrowed from a woodcutter and the fallen heroes were broken up like lobsters!

The needed resistance of armor was determined by "proof" and when it had been proved and marked, it was known as "proof armor." The test was the very practical one of shooting at the metal plates from a fixed distance with a crossbow of known strength. Where the bolt hit the armor a little key-shaped mark was stamped on the metal. For double proof which was tested at closer range, two marks were made.

Many suits of plate armor from this period and later are preserved in museums, and it's interesting to note in passing that nearly all of them are too small for a modern man to wear. There are exceptions: a suit in Windsor Castle which belonged to Henry the Eighth looks easily big enough to cover *two* men and a small boy.

70

Tournaments were now more popular than ever and a lot less dangerous. A heavy wooden barrier about three feet high was now built down the middle of the lists and each knight had to keep to his own side of it. This eliminated the collision of the horses and saved a lot of good horseflesh. The lances now were made of light, brittle wood which let them shatter easily, and the knights found that they were hurt less if they didn't try quite so hard to stay in the saddle. As an aid to dismounting, the cantles of the saddles were lowered so a man could slide over his horse's tail and clang down into the dust with no worse injury than a few bruises.

As the century progressed the barriers became higher, until they were so high that the knights could barely get a poke at each other across them. At the same time, the tournament helm reached its ultimate size and weighed about thirty pounds. The extremes of play-acting and display which went on in connection with tournaments at this time have no place in a book about weapons.

The English knightly swords changed not at all, but in France, toward about the middle of this century, the big slashing sword began to give way to a narrow blade which was exactly the right tool for slipping into the joints of armor. This sword was the predecessor of the rapier. When a

BILLMAN WEARING BRIGANDINE JACKET. Scales were riveted to the inside of the fabric.

striking weapon was needed, the French knight resorted to a clout with his mace.

Real war was becoming a grimmer business. Though in theory it was considered impolite for common soldiers to shoot at mounted gentlemen, in practice the common soldiers were doing more and more of it—and with relish. At Agincourt in 1415, the English longbow against great odds once more flattened the ironclad might of France and the prestige the archers gained began to extend to other classes of foot soldiers.

Infantry became important enough to wear armor. Nearly everybody had a headpiece of some kind and followed his fancy or his luck at looting, in the matter of body protection. Breastplates were now common and so were shirts of mail. The invention of wire drawing brought the cost of mail down. Brigandines, which were leather or cloth jackets closely studded with metal scales, became very popular; and for those who could afford nothing better, there were quilted jerkins.

The infantry was becoming divided into special-

HELM, FIFTEENTH CENTURY

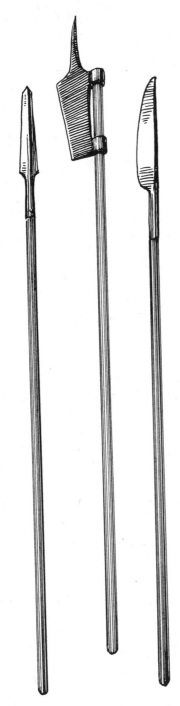

variety of heads. The three principal ones were the poleax, the oxtongue and the glave. Pictures are better than words to describe them. The poleax was literally a battle-ax on a long pole; the oxtongue was a spear with a two-edged blade for a head; and the glave was a big knife blade with a pole for a handle. Pole arms were useful against cavalry and in any close fight. One of their major jobs was protecting archers, arbalesters and gunners while they were reloading.

The military flail was also used by foot soldiers. It was simply two stout sticks swiveled end to end. One stick served as a handle while the other was thrashed about to do what damage it could. It seems, somehow, that if a man had to go into a fifteenth-century battle, he might wisely choose some other weapon than the flail.

Though longbows in the hands of experts could rule the field and guns were slowly improving, this was above all the century of the arbalest. When it was perfected, it provided a powerful bow which could be handled by common soldiers who had no special training and were not strong enough to draw a longbow. Its development happened something like this: Crossbows were made stiffer to increase their power; in the end they had to be made

**POLE ARMS**

*Oxtongue    Poleax    Glave*

HUNTER SHOOTING
AN ARBALEST

ist groups. Aside from the archers and culverin men, there were javelin throwers, sword-and-buckler men and men with pole arms. The Saxons had used their long-handled bills at Hastings. In time the hooked blade sprouted a couple of spikes and other long-handled weapons appeared with a of steel and looked not unlike one leaf of a wagon spring. Better gear than the simple little cord-and-pulley was needed to bend them. There were two solutions to the problem and both of them were in wide use for a long time. Because of the resemblance of some of this tackle to that used for setting a ballista, an English translation made this kind of crossbow an *arbalest*. The word doesn't get much use and its meaning is a bit faded; since all crossbows are not arbalests, but all arbalests *are* crossbows.

The earlier and more powerful of the two setting-rigs was an elaboration of the cord-and-pulley system. It used four pulleys working as two parallel purchases, and did the needed pulling with a windlass which slipped over the end of the stock and was turned by two hand cranks. It could set the heaviest of bows but took its time about it, and it was a constantly tangled mess to handle. Between shots it usually lay on the ground, and out of action it was a cumbersome burden on the archer's belt.

The other setting device was also clumsy and heavy but it was a little faster, and it had at least the advantage of having no strings to it. It was known as a *gaffle*, a *cric* or a *cranequin* and it worked on the rack-and-pinion principle. It operated in a "gear case" which had a loop on it to slip over the stock of the arbalest. On the case was a crank which turned a small pinion inside. In its simplest

GAFFLE FOR SETTING AN ARBALEST

SETTING A STRONG ARBALEST WITH WINDLASS AND TACKLE

THE "WORKS" OF A GAFFLE

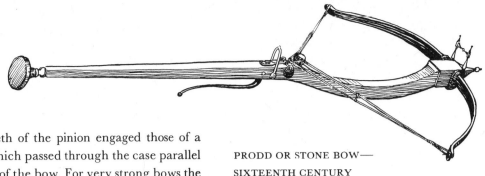

form the teeth of the pinion engaged those of a long rack which passed through the case parallel to the stock of the bow. For very strong bows the advantage was increased by putting a gear between the crank-pinion and a rack-pinion; this kind is shown in the drawing on page 73. The forward end of the rack had two hooks on it to hold the bowstring. The gaffle was disengaged from the string and removed from the stock before shooting. The same lock and nut which served the simple crossbow was used on the arbalest.

The maximum range of a strong arbalest in dry weather (the range was about zero in the rain) was up to a hundred and twenty yards. To get this much the bolt had to be shot upwards at an angle of forty-five degrees. Shot level at what's called point-blank range, a bolt would carry about sixty yards. For close fighting this was an effective weapon; it could pierce ordinary plate armor and inflict a bad wound.

In addition to their use in war, crossbows were excellent sporting weapons. For this purpose they were usually light and could be set with a lever called a "goat's foot." There was also a type of crossbow called a "prodd" or stonebow which was

PRODD OR STONE BOW—
SIXTEENTH CENTURY

light enough to set by hand. It fired clay or metal pellets and was especially valued for shooting birds. The later prodds had sights and double strings.

Arbalesters were generally put in the front ranks with the longbowmen behind them and shooting over their heads. The arbalester carried his own pavise strapped to his back, which he turned to the enemy while resetting his bow. His quarrels he carried in a bag at his belt and an additional supply followed the army in a cart. By and large, the arbalest seems to have been a better weapon for the defense than for the offense, though it was used for both. For loophole-shooting it was great, and loopholes came to be cross-shaped to give the arbalesters a wider range of vision. Three men working behind a loophole and shooting in turn could keep up a good steady fire. The little cubbyholes back of the slits in later castles are just big enough for three men.

In the second half of this century the culverin, though it still couldn't equal the longbow in the hands of an expert, began to crowd the arbalest in Europe. Its shape and balance were improved and it began to look more like a gun. The heavier culverins were still served by two men, the muzzle now resting on a forked stick. These heavier guns could be fired nearly as fast as an arbalest and they hit harder at greater distances, penetrating all but the very best armor at ranges up to a hundred and fifty feet.

The lighter culverins could be managed by one man alone. As the design improved the stock was curved downward somewhat and was given a broad butt which could be rested against the gunner's chest. Barrels began to be made longer which improved accuracy, though the best a man could do shooting from his breastbone was to let go in the

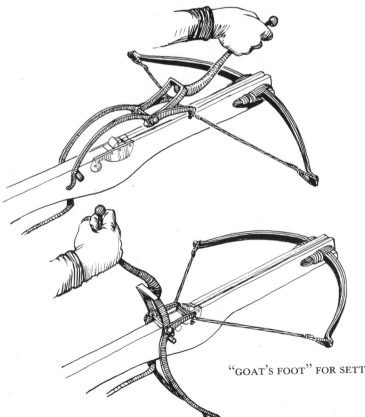

"GOAT'S FOOT" FOR SETTING LIGHT HUNTING BOWS

SETTING A PRODD

GAME BOLTS

WAR QUARRELS

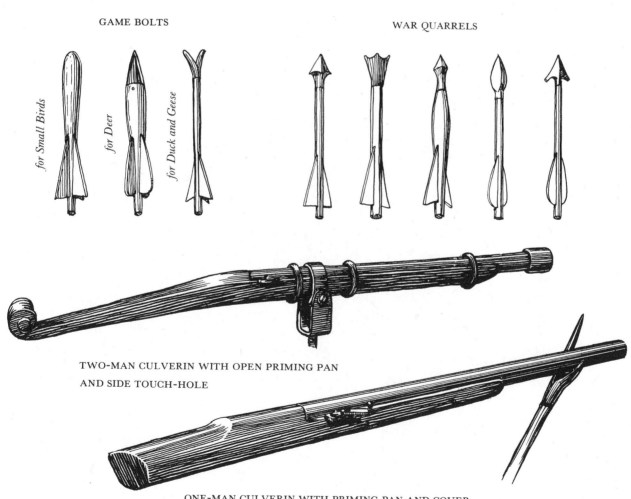

*for Small Birds*

*for Deer*

*for Duck and Geese*

TWO-MAN CULVERIN WITH OPEN PRIMING PAN
AND SIDE TOUCH-HOLE

ONE-MAN CULVERIN WITH PRIMING PAN AND COVER

general direction of the target. Firing was made easier by the addition of the priming pan which continued to be used on guns until the early nineteenth century. This was a little metal dish with a movable cover fastened to the gun barrel just below the opening of the touchhole, which was now drilled in the side of the barrel. Loose powder, ignited in this pan, would flash through the hole and set off the main charge.

The impure powder of early days would so foul up a hand-gun barrel that it had to be cleaned after every shot. This was the chief reason for the slow rate of fire from guns and the reason archers made jokes about gunners. The Germans then had an idea. They cut grooves on the inside of a gun barrel to give the dirt a place to accumulate, out of the way of the bullet. These grooves were just straight "ditches" from one end of the barrel to the other; years later they were given a spiral twist, called rifling, and served another purpose entirely. Before the fifteenth century was out the matchlock had been invented, but it was not of great importance at first.

The enormous bombards still threw stones. There was no way to allow the big guns to recoil when they were fired. They were fixed in a wooden frame strong enough to keep them where they were. This would have been possible only with the fifteenth century's weak powder.

The operation of large cannon was left in the hands of civilian experts for many years. They were highly paid professionals and as temperamental as emotional actresses. Like the Free Companions, they often changed sides without notice if they saw a chance to better themselves. Though the cannon could now surpass the trebuchet, it was much more costly and its rate of fire was much slower; two or three shots a day was good for a bombard. So the trebuchet remained in use and was not abandoned entirely for more than two hundred years. Cheap and handy, it was built a bit more carefully; otherwise there was no change in it.

Now, though it didn't know it, feudalism was done for. The baron in his castle was no longer safe from attack. Four or five hundred pounds of

LARGE SIEGE BOMBARD

stone plunked against his gate would knock it down, and even his thick walls could be crumbled by a bombard.

It was the Germans who finally managed to cast iron balls which would fit the bore of a cannon more tightly than stone ever did. This reduced the "windage" between the ball and the bore and used a lot more of the force of the explosion to push the projectile. About the same time purer saltpeter was produced and hence powder became much better. Stronger explosions more tightly confined were too much for the bombards; they blew up all over the place. When this danger appeared, gunners began to lay a train of powder from the touchhole along the top of the barrel all the way to the muzzle. Then the cannoneer lighted it and took himself hastily out of harm's way while the train was burning back to the touchhole.

About 1470 there was a quick trend back toward smaller cannons which could have thicker-walled barrels. After all a small iron ball with real kick behind it could do as much damage as a two-hundred-pound stone which just barely reached the target.

These smaller guns were real cannon. They fired a ball instead of merely tossing it. The bores ran from about two to four inches in diameter, and the barrels usually were long to take full advantage of the expansion of the gases of the explosion. Most of these guns were breechloaded into a hollow breechblock which was wedged in place against the breech. This didn't fit very tightly but it served at the time.

Some of the smaller guns were now cast in one piece and were mounted in frames so arranged that the angle of fire could be changed as needed. These frames had to be heavy enough to hold the gun in place when it fired. When guns go off they try to jump backwards. Modern guns are allowed to do it for a short distance and the recoil is absorbed and gradually slowed up, but in the fifteenth century all guns, even the biggest, were held rigid by timbers strong enough to take the shock of recoil.

Somebody in this century came up with the idea of holding a cannon and balancing it on a couple of lugs called *trunnions* which were cast as part of the barrel. With these the elevation of the shot

BREECHLOADER

77

could be controlled by resting the breech on a big wedge called a *quoin* which could be moved forward or back as needed.

The problem of firing red-hot shot was solved by putting a wad of damp clay between the powder and the heated cannon ball. The Germans also invented a bomb which was a hollow iron ball filled with gunpowder, but it took a century or so to learn how to shoot one from a gun; the first ones were just tossed by hand. Fire pots were tossed too. They were pierced iron balls filled with burning oil, gunpowder and powdered metal. They didn't explode, or at least they weren't supposed to; they merely spat flames from their holes. Tossing one must have called for some dexterity.

There's a weapon of this period of which few remember anything but its name—the petard. It had little to recommend it. In essence it was an iron bucket which was filled with gunpowder and hung on the gate of a stronghold to explode there and blow in the gate. Presumably the "gunner" had to drive his own nail to hang it on and bring along his own courage for the job.

## Matchlocks and Wheel Locks (1500—1600)

Because of the changes it forced on the ways of warfare and in spite of the awkwardness of the weapon itself, this was the day of the hand gun. It was the Spanish, coming into their spurt of military glory, who made the basic improvement. This was a way to make a gun shoot like an arbalest, by squeezing a trigger. In the late fifteenth century they invented the arquebus which was soon taken up by the rest of Europe. It was equipped with a matchlock. The Spaniards also discovered that they could sight a gun better and absorb its recoil better if they pressed its stock against their right shoulder to fire it, but the rest of Europe preferred for a while to go on shooting from its chest.

The stock of an arquebus was better shaped than that of the older culverin and its weight balanced better. The great advantage was in the means of firing—the matchlock itself. The touchhole and the priming pan with its cover weren't

ARQUEBUSIER AND HELPER

78

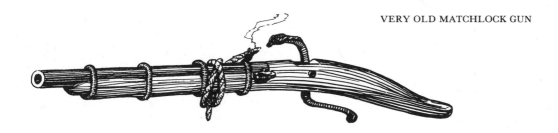

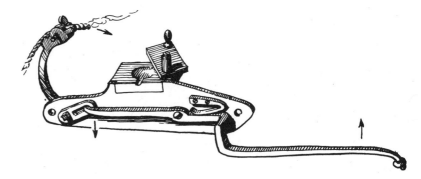

THE INSIDE OF A MATCHLOCK

new and the old slow-match was still used. The new thing was a movable clamp known as a *serpentine* which held the match on the gun. The serpentine was pivoted on a pin and connected to a big trigger like the one on an arbalest, so that it moved when the trigger was pressed and dipped its smoldering match into the powder in the pan, which had been uncovered by hand. The serpentine didn't snap forward like the hammer or bolt of a modern gun, it moved only as fast as its leverage on the trigger would move it. There was a spring to return it when the trigger was released. What was the advantage? The gun could be sighted and the arquebusier didn't have to take his eye off his target to look for the touchhole.

Shooting one of these things was a major operation. Before loading, the burning match was removed from the serpentine to avoid accident. Then coarse powder was measured and poured into the muzzle of the gun. Next the lead ball (usually cast by the soldier himself) was dropped in with a wad of rag rammed down on top of it to keep the ball from rolling out when the piece was leveled. Now the pan was uncovered and a little fine-grained priming powder was carefully poured into it, the cover being instantly closed again; any loose powder grains were meticulously blown off. The match burned at both ends to make sure of keeping fire on hand. One end of it had its ash knocked off and was carefully adjusted in the clamp of the serpentine, so that just enough of it stuck out and the coal was blown upon until it glowed nicely. A wait before firing would necessitate readjusting the match in the clamp. Now! Open pan cover. Squeeze trigger—and about half the time nothing happened except that "flash in

the pan" which we still talk about, or sometimes not even that.

If the gun did go off, the soldier set about the long job of cleaning up for the next shot. Powder was very dirty then, and even the touchhole had to be carefully cleaned out with a priming wire. Wet weather made matches go out and made powder useless. Gunpowder absorbs moisture quickly. It was not for nothing that Oliver Cromwell, whose men used guns like these, said, "Put your trust in God, my boys, but keep your powder dry."

The arquebusier carried a big flask of regular coarse gunpowder and a smaller one called a touchbox with the finer priming powder in it. The necks of these flasks held just enough powder for one charge. The soldier put his thumb over the top and turned the whole thing upside-down. A little cutoff was pressed open and the powder ran down and filled the neck. The cutoff closed itself by a spring when it was released and the charge could be poured in without spilling much powder. Extra match cord went around the hat (sometimes *in* the hat in wet weather) or around the left arm or hung in a bunch at the belt. Bullets were in a belt pouch with a couple for immediate use held in the mouth; soldiers who sur-

BULLET POUCH AND TOUCH-BOX

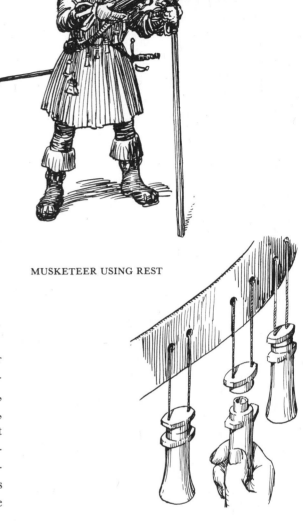

MUSKETEER USING REST

WOODEN POWDER CHARGERS ON A BANDOLIER

rendered with honor "marched out with their bullets in their mouths." In addition, the matchlock man carried a ramrod (in the gun stock), scrapers and bullet extracters and cleaning rags, bullet lead and a brass mold for casting it, flint and steel for relighting matches, and nearly always he wore a sword. In short, he was as encumbered as a modern doughfoot and needed his helper, who tended a small fire and carried some of the equipment.

The musket followed the arquebus out of Spain and was in fact only a larger and longer version of the same weapon, designed solely to puncture armor at long range. A musket was so heavy that it was necessary to fire it from a rest which the musketeer carried by a loop of cord around his wrist. The rest had a sharp end which could be used for defense between shots.

To speed up loading, powder charges came to be measured out in advance and were carried in little wooden bottles hung on a shoulder belt called a bandolier. Because of the weight and kick of a musket (it was charged with as much as two ounces of powder), all musketeers were strong, stocky men. Lead balls were the ordinary ammunition. A pound of lead made eight of them, if they were a tight fit, ten if they "rolled" in. This way of sizing by the number of bullets to the pound is the basis of modern shotgun "gauges."

In spite of having pioneered with cannon, the English were slow to take up small arms because for some time their longbow left little to be desired. When they did adopt the musket they made a naval weapon of it. For this job they frequently fired "sprights" which were extremely short arrows with vanes and *wooden* heads, but which nevertheless could be driven through the timbered sides of a ship. Two musket balls joined by a six-inch wire made a "chain shot." This would cut up sails and rigging better than a broad arrow but it did the musket no good whatever.

The caliver came along about the same time as the musket. At first the only difference between a caliver and an arquebus was that the caliver barrel was bored to a standard size so that, instead of each man having to cast his own bullets, a whole regiment could draw on a common stock cast ahead of time. Time changes meanings. After a while an arquebus was a heavy matchlock and a caliver was a lighter gun with a wheel lock, and the old word "culverin" became the name of a cannon of respectable size.

Obviously the need was for a gun which could make its own fire. The oldest one now known which accomplished the feat is the so-called "Monk's Gun" in Dresden. To us it looks more like a fire extinguisher than a gun, but it worked in spite of being very difficult to aim and hold. Its mechanism was copied not from a fire-fighting device but from a fire-lighting device which was in general use. A springy serpentine held a piece of iron pyrites close to the flash pan and pressing

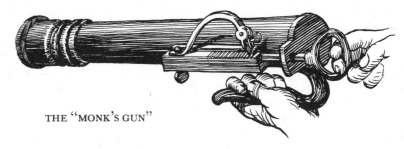

THE "MONK'S GUN"

**down** on the roughened surface of a flat steel bar. A loop handle served to draw the bar back smartly across the pyrites, automatically removing the pan cover with the same motion. A shower of sparks fell on the powder in the pan and produced the desired result.

The next step was to get this general principle into a more practical gun. The Nurembergers did it when they produced the wheel lock as early as 1515. It too had a serpentine with pyrites clamped in its jaws, but instead of a bar, it had a roughened steel wheel. This wheel was wound up like a clock with a "spanner" or key which put tension on a short chain attached to a strong spring. When the trigger released the wheel, it rotated rapidly against the stone and spurted the necessary sparks. The beginning of the rotation pushed back the pan cover which was then held open by a spring catch. The serpentine had to be moved by hand but this could be done any time after the gun was

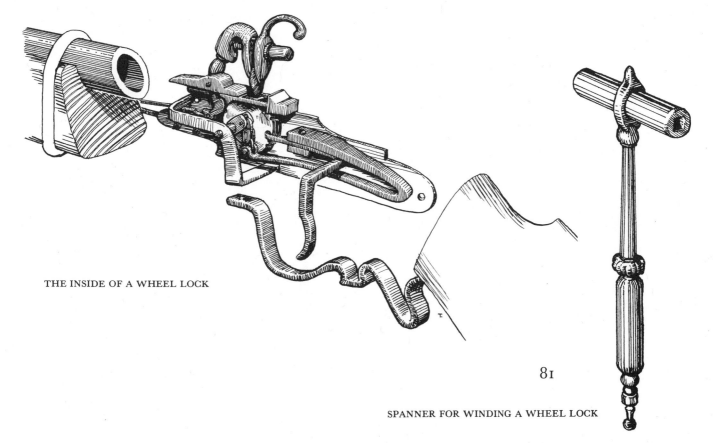

THE INSIDE OF A WHEEL LOCK

81

SPANNER FOR WINDING A WHEEL LOCK

loaded, and a spring kept it pressing on the pan cover until pulling the trigger moved the cover out of the way. The top of the wheel met the pyrites in the pan itself.

The wheel lock was excellent and very dependable, but it was also very expensive and its great cost kept the matchlock in use for a great many years. With the new lock the gun was for the first time a really good sporting weapon; it could now be loaded and "spanned," then held ready for instant use while the game was stalked. Sporting wheel locks for small game were usually made with very small stocks shaped to be held against the right cheek for firing; in fact, it's impossible to line up the sights of a wheel lock with the eye unless it is so held. The pistol was born with the wheel lock; this delighted the cavalry who hadn't made out very well trying to shoot matchlocks on horseback.

The early wheel-lock pistols were made with heavy balls on the butts of their only slightly curved stocks. In a pinch they could be reversed and used as maces. This ball was a pistol feature for quite a while but it came to be made a mere ornament with a neck too weak to be any good in a fight. Cavalry pistols were about two feet long, but smaller ones called daggs were soon made and soon outlawed because of the company they naturally kept. A statute was enacted in England prohibiting all guns shorter than three feet.

Of course somebody soon tried for a gun which would fire more than just one shot without reloading. The best ones had two or three barrels on one stock like a modern double-barreled shotgun; but there was also an alarming type in which as many as eight loads, with leather wads between them, were put into a single barrel and hopefully fired one at a time with a sliding serpentine and a series of touchholes. This never became very popular. An arquebus with four chambers, each of which could be loaded separately and rotated to line up with the barrel, proved to be too tricky also and to leak too much gas. A kind of "machine gun" was made by mounting a number of barrels on a wheel and bringing one after another into firing position; one man could load, another fire, and so on around the wheel, a man for each necessary operation.

When a gun misfired a man might find himself in a tough spot, and guns missed fire often. This led to odd combinations of weapons built with the idea of giving the gunner a second chance. In addition to the ball-butt, pistols often had dagger blades on them; battle-axes and even whips had guns in their handles. But the prize of the lot was the "holy-water sprinkle," a great spiked mace which had four or five small barrels drilled into its business end. Each barrel was loaded separately with bullets, but the powder went into a common

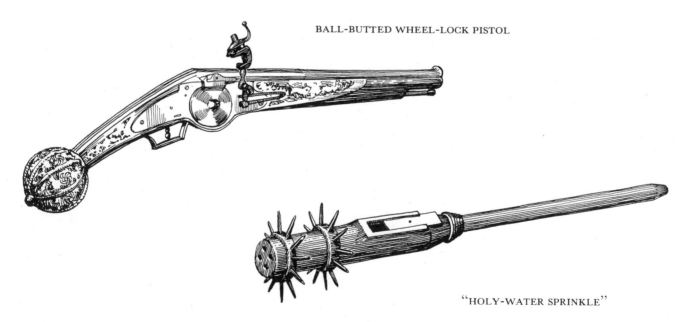

BALL-BUTTED WHEEL-LOCK PISTOL

"HOLY-WATER SPRINKLE"

chamber in the handle into which all of the breeches opened. It's hard to imagine a more "un-English" weapon but it was a great favorite in England in the sixteenth century.

The system of putting spiral grooves in a barrel caused the bullet to spin around an axis parallel to its line of flight, and hence to stick to that line of flight instead of tumbling this way and that as a ball did from a smooth barrel. Rifled guns came into general use on the Continent in the sixteenth century and were especially prized for hunting large game. The English stuck to the smoothbore for all uses and never considered the rifle as a military weapon until long after their unpleasant encounter with it in America.

## Soldiers (1500–1600)

Armored knights still operated through the sixteenth century but their effectiveness as fighting men dwindled steadily. Infantry was taking over and often the knights themselves condescended to do battle on foot. By 1600 they no longer took to the field as knights at all and in 1605 the publication of *Don Quixote* laughed them out of existence.

The harder-hitting bullets of the arquebuses required that armor be made hopelessly heavy and the introduction of the musket made any armor a man could carry almost useless. Ordinary soldiers hated it. For one thing, its cost was deducted from their pay; for another, they had to wear it. By the age of thirty some men were permanently deformed from the weight of the stuff.

This is not to say that armor disappeared in the sixteenth century, because it didn't. Maximilian of Germany invented a new kind, built with ribbed bulges, which to his eyes was handsome but which reminds us today of nothing so much as the old pot-bellied stove in the general store. The shapes used in Maximilian armor were copied from the puffs and slashes which were fashionable in civilian clothes at the time. Later its extremes were modified but armor, like the feudalism to which it belonged, had lost its usefulness in the world. Both still existed but only because men hadn't yet realized how useless they were.

KNIGHT IN MAXIMILIAN ARMOR

Other kinds of mounted soldiers had become common. When a knight sallied forth he now took along an assistant, also mounted, who was armed with a sword and javelins and was known as a *custrel*. Mounted arquebusiers appeared who fought with short-barreled matchlocks which a later age would have called musketoons or carbines. These men wore armor and helmets without visors because it was found that a gun couldn't be aimed very well through a peephole.

The invention of the wheel-lock pistol produced a mounted "pistolier." Since a pistol could be discharged with one hand, he carried a pair of them. The model on which all other pistoliers formed themselves was the German *reiter*. He was a member of a Free Company which rented itself to any army and generally he was an extremely unpleasant character. Mounted on big horses and clad in black armor to appear more fearsome, the reiters massed themselves for battle twenty ranks deep. Their fighting maneuver was the *caracole* which consisted of charging at the enemy one rank at a time, discharging their big, ball-butted weapons and turning aside to clear the way for the following rank. By the time twenty ranks had charged, the first rank had reloaded, spanned their pistols and were ready to have another go at it.

REITER

On the Continent guns replaced crossbows entirely early in the century, the English abandoned the arbalest in 1535 and after 1550 the longbow began to lose out to powder. Even in the latter half of the century however, an expert archer could outshoot any gunner in accuracy and in the number of shots he could fire in a hurry. It was that word "expert" that whipped the bow. Any soldier could learn to handle a gun satisfactorily in a few weeks but only a lifetime of training could make an archer. In old days kids began shooting as soon as they could hold a little bow, but that kind of enthusiasm had died and, too, powder was proving stronger than yew.

You'll have gathered that a soldier took some little time to prepare his gun to fire and that while he was about this he was practically defenseless. That's where the pikemen came in. Their enormously long spears, as long as any used in a Greek phalanx, were very effective against cavalry charges and a hedgehog of them proved to be the ideal cover for a company of arquebusiers. As long as the matchlock was in use armies held to a fixed proportion of pikemen to gunners, and the match-

84

PIKEMAN

lock wasn't abandoned until the beginning of the eighteenth century.

The pike evolved either from the sharpened stake which archers stuck into the ground in front of them, or from the similar use of a knightly lance. Pikes were as long as twenty-two feet, and this great length was the cause of an odd annoyance which afflicted their users. They collected rain and dribbled it down their staves onto the shoulders of men carrying them. Sometimes "handles" and tassels of various kinds were put on them as rain spouts. The great length of the pikestaff made it the favorite support for the severed heads of traitors when they were displayed in public to discourage imitation. This same length also made a pikestaff very noticeable, so whatever was obvious became "as plain as a pikestaff" and still is.

After the gunners and the pikemen had softened up the enemy resistance, pole arms which were shorter, more versatile and infinitely more personal were used for mopping up. It wouldn't be safe to say that the old shapes were improved; rather they were elaborated and some were given new names. A bill with two or three extra spikes on it was called a *guisarme.* The oxtongue sprouted a couple of curved points at its base and was known as a *partisan,* which was a thrusting weapon only. The glave grew a much bigger blade, with a few spurs on its unsharpened side, and was then called a *fauchard.* The poleax, with its spear part made much longer and its blade crescent-shaped, became a *halberd.* The variations of shape in these main kinds were wide and many of them were incrusted with ornamentation.

As the Germans were addicted to pistols, and the English were skilled with bows, so pikes and halberds were the weapons of the Swiss. After using them with conspicuous success in their own fights, they formed into mercenary companies and hired themselves out for the work, with always the reservation in their contract that they wouldn't be compelled to join in storming any strong fort. They seem to have had an ingrained fear of cannon—and of little else.

Of all the Free Companies only the Swiss have left a remnant to our own day. The Pope still

SWISS HALBERDIER OF THE VATICAN GUARD

maintains his Swiss guard of halberdiers who continue to wear the fancy costume and the steel body armor and morion of their greater days. This morion was the favored headpiece of the Spanish *conquistadores* in Mexico and Peru. It was a bowl-shaped iron hat which dipped on the sides and turned up in points in front and behind; nearly all of them had high circular "combs" across the crown, also running from front to back.

As was the case with the longbow, it took a strong man to wield an *espadon* or two-handed sword. Though the word sounds Italian, the weapon was most popular with the Swiss, the Germans and the Scots, whose "claymore" was originally double-handed. These long-bladed, long-handled slashers saw more use in this period than at any other time in history. Knights dueled with them on horseback in tournaments, and specialist foot soldiers swung them in actual war. They all had two-edged blades, most of which were straight, but there was a variety with a wavy blade, known as a *flamberge*. The advantage of the waves is not quite clear.

Both kinds were too long to be hung from the belt in a scabbard. Instead they were carried hanging down the back or shouldered as a modern soldier carries his rifle. To make this possible with-

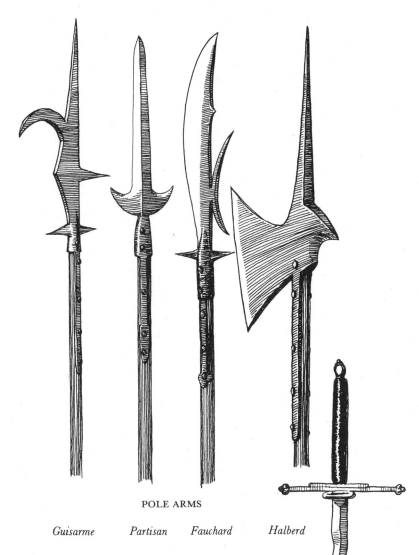

POLE ARMS

*Guisarme*     *Partisan*     *Fauchard*     *Halberd*

out slicing into the shoulder muscle, the blade near the hilt was covered with leather. It was the great length of blade which made these swords effective slashing weapons, and the same feature in the end caused them to be abandoned. A man swinging one of them needed plenty of space all around him, and he couldn't always get it. Once in a while somebody forgot to look, wound up for his swing and clipped the file behind him.

By 1500 the old knightly sword with its bulky, rigid blade had been pretty well replaced by lighter, thinner swords of the rapier kind and by shorter, very heavy-bladed swords like the Italian *Malchus* and the *cinquedea*. Not all thin swords were rapiers. The German *estoc* was a rigid thrusting sword with a grooved blade and a simple hilt.

85

FLAMBERGE

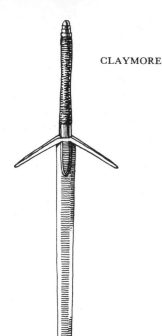

CLAYMORE

DUEL WITH SWORDS AND DAGGERS

DUEL IN THE FRENCH STYLE

made their own. Honor was very tetchy in those days; so much so that it was evidently difficult to speak to a French gentleman civilly enough to avoid provoking him to combat. This must have been true because in ten years of the reign of Henry IV of France six thousand nobles were killed in duels and thousands more followed them in later reigns.

Most sixteenth-century rapier hilts had, in addition to the simple grip and *quillons* as the cross guards are called, a complex guard formed of and around a series of metal rings. A clever swordsman would make use of this guard to disarm an opponent. Just at the end of the century the rings were filled in and shaped into a hand-protecting cup surrounding the blade, just below the quillons.

Anyone whose social position allowed him to carry a sword wore it in public at all times. At first rapiers were carried hanging down the back as the espadons were sometimes carried, but the hilt wasn't readily accessible there and it was difficult to be quick on the draw. Besides, it didn't look very dressy. Both problems were solved by a *baldric,* a wide belt hung over the right shoulder with its two ends attached to the scabbard in such a way that the sword hung diagonally across the lower back with its hilt at the left hip. Here it was an eternal nuisance to its wearer and everybody else; the forty-inch blade was forever whanging into things and tripping people.

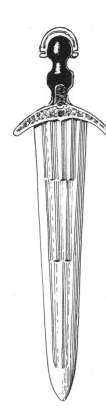

Rapiers usually had flexible, highly tempered blades and were given handsome and elaborate hilts. As in Roman times, the best blades were made in Spain at Toledo. Dueling with rapiers was very common in the sixteenth century. It replaced the old judicial combat as a means of settling arguments but lacked the same legal sanction. The system of attack and defense with thrusting swords which is called fencing originated in Italy, where at first the fighting was done with a sword in the right hand and a specially designed dagger in the left.

The use of the *main gauche* as the dagger was called, was gradually dropped when the fine art of dueling was taken over by the French and

CINQUEDEA

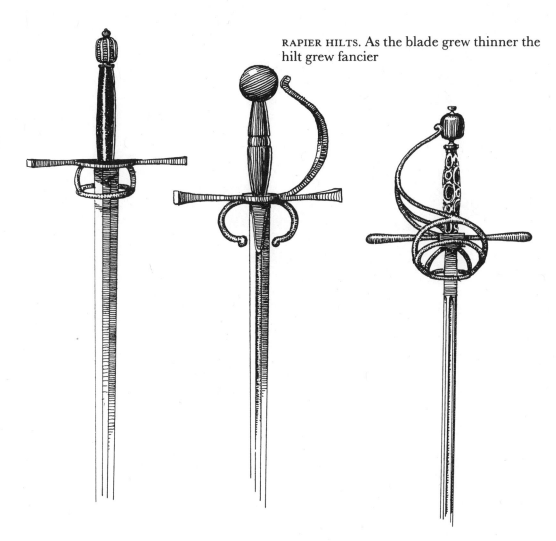

RAPIER HILTS. As the blade grew thinner the hilt grew fancier

SIXTEENTH-CENTURY HEADPIECES

NOBLEMAN WITH RAPIER IN BALDRIC

*Burganet*

*Morion-cabasset*

*Morion*

*Cabasset*

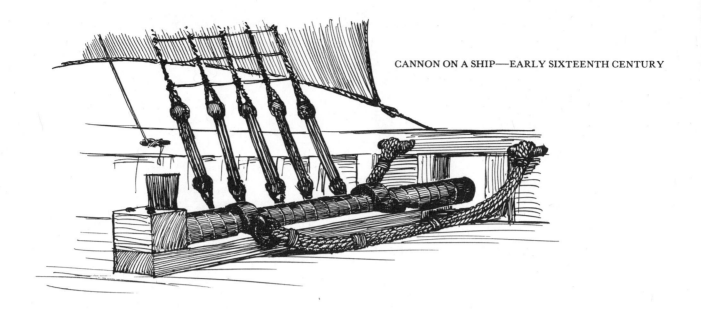

## Cannons (1500—1600)

No gun shoots better than its powder. The first bombards were fired with finely ground "meal" powder made of saltpeter, charcoal and sulphur mixed in equal parts, which resulted in a very weak, smoky explosive. When this was rammed into a gun, its fineness caused it to pack into something like a solid cake so that, instead of going off in an instantaneous flash when it was ignited, it merely caught fire at one end and burned through to the other, finally generating enough gas pressure to push its projectile out of the bore.

In the early days of the sixteenth century the bombards were still throwing stones at walled towns. For longer shots breechloaders fired cast metal balls, but instead of loading powder and shot into a recess in the breechblock, these were now put into the gun itself and somewhat better methods of keeping the block in place were invented. One of these was a kind of locking pin and another was a crude form of the interrupted screw which is the principal means of breech-locking in modern guns.

Cannons used on ships were not different from land guns and like them were rigidly mounted in strong wooden frames. There is some indication that while the frame of a land gun was staked down to be immovable, those on shipboard may have been allowed to slide back across the deck

a short distance when they recoiled. This relieved the strain on the ship's timbers. The backward motion was checked by heavy ropes secured to the ship's bulwarks. In any case, the guns were so enormously heavy that the *Mary Rose* of Henry VIII's time sank under the weight of hers. Some of them were recovered from her hulk three hundred years later and are preserved in the Tower of London.

About 1520 a trick was discovered which added punch to gunpowder without making any change at all in the proportions of the stuff. Instead of the "meal" consistency, powder was "corned" into coarse grains so that even when it was rammed, little air spaces remained in it. Fire could travel through these and ignite the whole charge quickly and uniformly. It was from this point on that cannons ceased being substitutes for siege engines and became *artillery*.

The art of casting and boring large guns improved. Breech-loading began to disappear because no known way of locking a breechblock was tight enough to hold the more violent expansion of corned powder. Cannon continued to be loaded from their muzzle ends until the latter part of the nineteenth century when properly forged and machined steel barrels were developed. Most of the early cast cannon were bronze be-

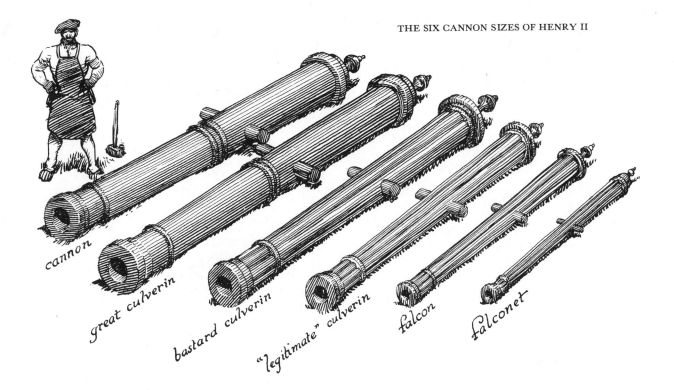

cannon

great culverin

bastard culverin

"legitimate" culverin

falcon

falconet

cause it was an easy metal to work and, being less brittle than iron, was correspondingly less likely to break when overcharged. Some bronze ones did burst, however, because proportioning the metals in the alloy was largely guesswork—maybe it was right, maybe it wasn't—and if it was right you didn't know why and had no way of repeating it.

Cannon were made in all sorts of lengths and diameters of bore according to the fancy of the king or the founder, as well as to fit them for some special purpose. By and large, the longer the barrel the longer the range, because the ball stayed in the bore until the powder burned completely and put the full expansion of its gases behind the shot.

By mid-century there were so many sizes of guns that it became necessary to bring order out of the confusion by assigning them to named classes. Henry II of France cut his down to six sizes but few went that far with simplification. The Spanish used twelve sizes, the English sixteen, ranging from the *cannon royal* which weighed four tons and fired a seventy-four-pound shot, down to the little rabinet which, though it weighed

three hundred pounds, had a bore little larger than a musket and fired a five-ounce ball.

We use the word cannon for all large artillery, but in the sixteenth century a "cannon" was a gun of a definite size and type, blood brother to the bombard and used mainly as a siege weapon to break down walls. The long-range guns borrowed the name of a small-size ancestor and were called culverins. A sixteenth-century culverin might have a barrel eleven feet long; its bore would be about five inches in diameter; it would weigh more than two tons and, using twelve pounds of powder, it could throw an eighteen-pound shot more than five thousand yards.

You'll see from this that corned powder plus better casting had advanced the gentle art of bombardment by a great leap. After this there was no really radical change in the muzzle-loading gun for three hundred years. In time a better system of standardizing sizes was found and fewer sizes were used. Here are the names of the artillery pieces used by the English in the latter half of the sixteenth century, arranged in order according to size of bore and beginning with the largest: the *cannon royal, cannon, cannon serpentine,*

MOVING A CULVERIN

*bastard cannon, demicannon, pedrero* (this threw six-inch stones), *culverin, basilisk, demiculverin, bastard culverin, saker* (a handy six-pounder), *minion, falcon, falconet, serpentine, rabinet.*

The later bombards had been moved around on four-wheeled trucks and some of the smaller ones may have been mounted permanently on wheels. About the middle of this century guns began to be given two-wheeled carriages. These seem to have been used even for the largest cannon which required as many as forty horses to move them. Guns cast with trunnions were mounted on them so that the barrel could be raised and lowered, those with no trunnions were simply cradled in frames. A heavy wooden *trail* projected from the carriage behind the gun. One end of it rested on the ground for firing. On the move the horses were hitched to it, the gun being towed backwards. The trail of later guns was supported for moving by a cart called a *limber,* but in the sixteenth century the gun was so mounted that it balanced on its wheels.

The Germans invented mortars which were very thick-walled short guns built to drop shot on an enemy by throwing it upwards at a very steep angle. Most mortars were chambered, that is, their powder was put into a recess at the back of the bore which was smaller than the bore itself. Some long guns were made that way too, with the idea of gaining a thick wall at the point where the explosion took place; it made long guns hard to load.

It was the Dutch who learned to shoot bombs from mortars. A bomb was a hollow metal ball filled with powder and having a small hole in it for a fuse. First they tried "single firing" which was putting the bomb into the mortar with the fuse down, in contact with the propelling charge. That didn't work. Firing the mortar often drove the fuse right into the bomb and it blew up right in front of the gun. Then they tried "double firing" with the bomb turned over, fuse up, and the gunner lighted the fuse by hand at the same time he lighted the touchhole of the piece. This required a nice sense of timing and a state of mind prepared for all eventualities, since guns often missed fire and a lighted bomb in a mortar barrel with nothing to push it out, could lead to trouble. It wasn't until 1650 that someone discovered, probably by accident, that double firing was unnecessary. The heat of firing would light the fuse even though it was turned *away* from the exploding charge.

"DOUBLE FIRING" FROM A MORTAR

90

## Cavaliers and Snaphances
## (1600—1700)

War began to get tougher. Men thought somewhat less of glory, chivalry and heraldry and more of defeating the enemy by any possible means. Armor, having reached its greatest bearable weight, still wouldn't stop musket bullets. If it gave no protection, why wear the stuff at all? Complete armor was still considered indispensable for parades and was worn at sieges. It was replaced in ordinary fighting by three-quarter armor and half-armor. Three-quarter armor stopped at the knees and was worn by heavy cavalry. Half-armor was nothing more than a breastplate and a backplate worn with some kind of iron hat. The Spanish and Portuguese clung to armor longest; they seemed to mind the heat of it less than the northern soldiers did. Vestiges of armor clung to military uniforms as ornamental

tinware long after it was useless. The Count de Rochambeau, when he arrived in America to take part in the Revolution, was described as being "in shining armor dressed"; and the earliest portrait of George Washington shows him wearing a silly little iron collar.

English and French pikemen wore some armor until they were abolished about 1675. By then other soldiers had followed the example of the Swedes and cut their metal down to a single cuirass (breastplate). Some classes of soldiers clung to this more or less permanently for dress uniforms, and it may still be seen encasing the Horse Guards on duty in London.

Dragoons were invented at about this time and they wore cuirasses. They were mounted gunners but they dismounted to shoot. Ten men would go

91

into action while an eleventh held their horses. One reason for this was the extreme difficulty of handling a matchlock gun on horseback. Because it was a nuisance even to carry a burning match in the saddle, most dragoons soon changed over to the short wheel locks called "dragons" which gave them their name. Later they carried flintlock musketoons, which were short muskets.

The cavalier who appeared early in the seventeenth century owed much to the reiter. He was a mounted pistolier, a gentleman soldier serving as a replacement for the now useless knight. A very dashing character, much given to colored sashes, leather "buff coats" and wide, plumed hats, the cavalier nevertheless usually wore three-quarter armor in battle, with boots on his lower legs and spurs at his heels.

The cavalier's sword was a rapier, a "bodkin" in his words; his four eighteen-inch pistols were wheel locks, more often called firelocks in their own day. These were no popguns. A pound of lead made them twenty bullets for a tight fit, twenty-four "roweling in." Since a man can't very well shoot more than two pistols at once, the cavalier was attended by a boy on a nag to carry the spare arms and a sack of oats; obviously a vestige of the squire who followed his knight.

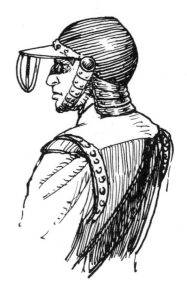

During the Puritan Revolution in England the Royalists were all called "Cavaliers" because most of the gentry were on the King's side, but the cavalier as a type was by no means exclusively English. By the way, Cromwell's Puritan soldiers were known as "Roundheads" because of the dome-shaped salades they wore.

By 1630 arquebuses and calivers were gone completely as military weapons because they couldn't compete with muskets in hitting power. Muskets had been made gradually lighter and could now be fired from the shoulder without a rest to hold up the barrel. They continued to be fired by matchlocks through most of this century, not because the matchlocks were very good, nor because there was nothing better for the purpose, but simply because they were cheap. Wheel locks were now made everywhere and they had reached a high degree of mechanical excellence, but they cost fancy money. The first crude flintlock had been invented in the later years of the sixteenth century, but it wasn't too dependable and it, too, cost considerably more to make than a simple trigger-and-serpentine.

This first flintlock was called a *snaphance* (originally spelled "snaphaunce"). The word and the lock were both originally Dutch, and the word described a pecking hen as well as the pecking action of the lock. There's an old story that says

CAVALIER AND ATTENDANT

92

the device was invented by poultry snatchers who couldn't afford wheel locks and who dared not carry matchlocks because the glowing matches made them targets in the dark. This is an unlikely yarn. More probably the snaphance was named in derision by the wheel-lock makers whose business was threatened by it, or simply in a burst of descriptive originality by the people who used it.

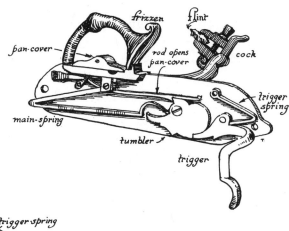

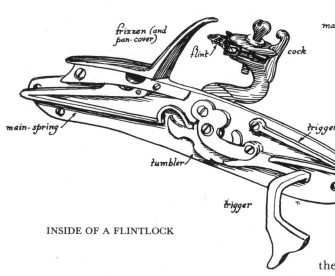

INSIDE OF A FLINTLOCK

Sparks struck from flint-and-steel had been used to start household fires from the time the wood-friction method went out of style. All that was needed was a mechanism to adapt the principle to a gun and that had been partly solved by the design of the wheel lock. From it was taken the priming pan and its cover and the "cock" into which the flint was clamped. The new features were: a main spring roughly the shape of an open safety pin, which would throw the cock smartly forward and down when it was released; a catch which, in the earliest ones, held the cock back until the trigger was pressed by projecting through the lock plate to engage a notch in the cock itself; and a movable steel frizzen against which the flint could strike to make sparks. The frizzen also had a spring and catch on the outside of the lock plate which held it in an "up" position for priming and in a "down" position for firing. It had to be moved by hand.

The gun having been loaded from the muzzle, firing procedure went something like this: with the frizzen in its up position, the pan was primed and its cover was slid back over the priming powder; the cock was drawn back by hand until the catch clicked into its notch and held it; the frizzen was then moved down until its lower edge was immediately above the flash pan. Pulling the trigger released the cock which snapped forward, striking its flint with a scraping motion against the face plate of the frizzen and showering sparks downward into the priming pan. By the time the flint met the steel the pan cover had been slid open by a link connected to the trigger mechanism, so the powder was exposed and ready to be ignited.

The improvements which converted the snaphance into what is called a flintlock weren't long in coming. England and France have had some difference of opinion as to which first introduced them around 1625. One minor one was the half-cock safety, a catch which held the cock halfway back in such a way that the trigger couldn't accidentally be pulled; but the major one was the combination of the frizzen with the pan cover.*

* This combination of cover and frizzen was often called the "hammer" but since the hammer on later guns was quite a different thing, it seems best to avoid confusion by skipping the term here.

93

The cover was hinged horizontally at its forward edge and the curved "steel" projected upwards from its rear edge. When the trigger was pulled, the flint striking the curve of the frizzen, flipped the cover open and at the same time showered sparks on the priming powder.

Though details varied and workmanship improved with time, this was the way all flintlocks worked and the way they *still* work, for many a one is in active use in the hinterlands of the world, and the little town of Brandon in England does a considerable business in "knapping" flints for them.

The *fusil,* as the flintlock gun came to be called, wasn't adopted immediately by armies—partly because of cost, but mostly because it missed fire too frequently. Good gunsmithing improved this failing but was never able to eliminate it.

The first English military flintlock gun was introduced in 1682. The famous "Brown Bess" soon followed and was still shooting a hundred and sixty years later. Brown Bess got her name from her color; not only was her walnut stock naturally

brown but her barrel was artificially browned by rusting with acid and rubbing down. She had no rifling, the inside of her barrel was a smooth tube and she was known as a smoothbore musket. For rapid loading, the lead ball used in her was small enough to drop into the barrel, fitting almost loosely enough to rattle. When this was fired it went whim-whamming off on an erratic course in the general direction of the target. Even in the hands of a good marksman Brown Bess wouldn't "hit the broad side of a barn." This was not considered important because the military theory of the time valued not marksmanship but volume of fire. There must have been something to it because Brown Bess did all right.

Even the earliest of these muskets was equipped with a bayonet which slipped on to the muzzle end of the barrel so that the fusileer could be his own pikeman. That came about like this: Back in the second quarter of the seventeenth century a posse of French peasants, out hunting bandits, ran out of powder. In desperation they jammed their knife hilts into their gun muzzles to convert the arquebuses into spears of a sort. The idea caught on, and in 1649 someone in Bayonne started manufacturing blades which were hafted with tapered plugs *intended* to be stuck into gun muzzles. In a surprisingly short time the advantages of this were impressed on European generals and the plug *bayonette* put the pike entirely out of business. Then in William III's time, some Englishmen were completely flabbergasted when the French charged them with fixed bayonets and paused to fire their guns *without removing the blades* from the barrels! The plug had been replaced by a couple of rings which slipped over the barrel. These rings were soon superseded by a tubular socket.

There was another flintlock weapon which was introduced into England from Holland about the middle of the seventeenth century and which became the Chief Defense of the English Home. This was the blunderbuss. It had a short, thick barrel with a trumpet-shaped muzzle for easy loading. The barrel was usually brass and fired a couple of ounces of slugs chopped out of sheet lead.

This was the period when coaches were coming into use in England, and with them came the dubiously romantic highwayman. A blunderbuss

PLUG BAYONET

SOLDIER WITH A FLINTLOCK MUSKET, LATE SEVENTEENTH CENTURY

94

BRASS-BARRELED BLUNDERBUSS

COACHMAN WITH A BLUNDERBUSS

was murderous at close range and one of them, lying across the knees of the guard or on the dickey-seat beside the coachman, made a handy defense against Dick Turpin. A blunderbuss also made a good weapon to keep over a cottage mantel to cope with those "things which go *bump* in the night." There came a time in the eighteenth century when such a gun was nearly as much a part of an English household as the teakettle.

Larger guns of the same type, mounted on swivels and throwing as much as a pound of slugs were used by the Navy and came to be known as "boat guns." Pot hunters, who slaughtered wild ducks and geese for the market, also swivel-mounted huge "punt guns" on their boats, but these had long barrels. Similar guns were used later in America; some of them had ten-foot gas pipe barrels with a three-inch bore and were loaded with a couple of pounds of small nails. They were illegal of course.

There was another terrible gun of the blunderbuss type which was used against poachers in England. This was the "spring gun." It, too, was swiveled, but it was set up in a wood with three or four long wires strung from it in several directions, just above the ground. When an unlucky poacher tripped on one of the wires the strain would instantly swing the gun around and fire it *along* that wire.

SPRING-GUN FOR DISCOURAGING POACHERS

## Field Guns and Bastioned Forts (1600—1700)

The one great difference between the cannon of the seventeenth century and those of the sixteenth was in the matter of weight. Powder was much improved. Not only was it corned but the proportions of the mixture itself were better, and a pound of powder would now do what once had needed three pounds. Metal working improved too. Take the thirty-pounder culverin for instance: In Queen Elizabeth's army it weighed more than two tons, in Charles II's, though still firing the same size shot and throwing it just as far, it weighed only 1400 pounds—and it was a stronger gun.

In the field, lightness meant mobility and that has always been an advantage to any army. The great Swede, Gustavus Adolphus, made real use of light cannon and mobility. He invented a field gun which a couple of horses or a few men could easily put where it was needed. And he used his field guns directly against soldiers—to outshoot muskets.

His first try was the so-called "leathern gun," which actually had a copper barrel hooped with iron and covered with varnished leather. It was light and very handy but it wouldn't stand a big enough charge to shoot far, so Gustavus shifted to cast iron, but he kept his guns light; his four-pounder weighed five hundred pounds. He trained three-man gun crews and invented a cartridge which contained both powder and shot to speed up loading.

Before his time an army dragged along one cannon for every thousand men. Gustavus provided six of his nine-pounder, demiculverins for each thousand and, in addition, gave every regiment two of the four-pounders. It was he who standardized the sizes of cannon by their shot weights and bore diameters. He also used *case shot* or *canister* against infantry. Originally this idea was carried out by firing a basket of small stones from a bombard or a big mortar. From his field guns Gustavus fired tin cans filled with musket balls or scrap metal to scatter among enemy troops. Canister was replaced later by shrapnel but it has now come into its own again and shrapnel has been put on the shelf.

CANISTER

Siege weapons were progressing. From Holland came the coehoorn mortar which everybody adopted. The coehorn, as the English called it, was usually small. The barrel was about twice as long as the bore diameter. About this time the discovery was made that a bomb fuse turned towards the muzzle of a mortar would still be ignited by the blast of the discharge, to the vast relief of all bombardiers. Once this fact was established, the idea of shooting a bomb from something with greater range than a mortar presented itself. The short cannon, mounted to shoot at a fairly high angle, which was the result of this idea, was called a *howitzer*. Guns called howitzers are still used, but the distinction between them and other artillery

96    SWEDISH CAST-IRON FOUR-POUNDER

SMALL COEHORN MORTAR

pieces has faded. Today almost any gun can be fired at a high angle if necessary and all of them fire explosive shells.

Another radical idea then presented itself: Why not aim the gun with the deliberate idea of *hitting* a target? Gunners began to try to control the range of their pieces by measuring the angle of elevation. The long leg of a "gunner's quadrant" was put into the bore and the angle read from the position of a plumb line on a protractor scale. The theory was that a gun would shoot ten times as far at an angle of forty-five degrees as it would horizontally, or "pointblank." Cannon had no sights at this time nor for a long time afterwards, but a good gunner would use his "level" to mark a point on the muzzle and one on the breech, both (he hoped) exactly over the center line of the bore. Then he would connect the two with a chalk line and sight the gun by that. No explanation is given of why nobody thought of scratching a permanent line on the barrel. One difficulty of sighting a cannon in those days was due to the fact that the barrel was smaller outside at the muzzle than at the breech. Allowance-by-guess had to be made for the fact that a line sighted across the two points would elevate the bore considerably above the target and cause the gun to overshoot. Usually the first shot was deliberately fired short as a kind of measuring stick for the range.

There was plenty of shooting done; one war seemed to lead to another. Men fought over re-

HOWITZER (the carriage is a reconstruction)

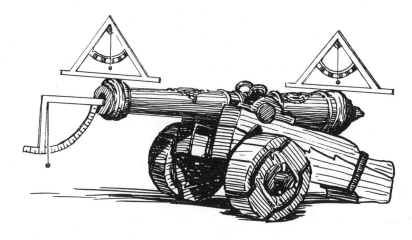

GUNNER'S QUADRANT AND LEVEL IN POSITION FOR USE

ligion, they fought over the shaky claims of one king to another's throne, they fought to escape from bondage and to subject other nations to bondage, much as men do today.

What were the guns shooting *at?* Men, as Gustavus Adolphus had taught them to do; ships, when the shooting was at sea; forts, where they defended cities and frontiers. Some forts were masonry and stood as proud towers above the sea like those the Spanish built in a chain through the West Indies, but these were put there in the sixteenth century and they were all but obsolete when they were built. Unless they were banked outside with earth, stone walls were no good against cannon fire. A gunner could just keep on pecking away at the base and presently the whole wall would come down.

The Italians were the first to experiment with low forts specially planned to absorb cannon balls. These were given earth-banked walls. The French took hold of the idea and ran away with it. In the late seventeenth century they produced a genius at the art of fortification, His name was Vauban,

and some of the principles he developed are incorporated in modern defenses.

Vauban used the *bastion* system of constructing forts. A bastion was a projection from the main body of the fort which served the same purpose as a tower on a medieval castle wall: it allowed cannon to be so mounted that their fire would protect the straight *curtain* walls between the bastions. The whole theory of fortification was that each part should protect and be protected by another part of the structure. The ground plan of Vauban's forts was usually a regular polygon, and with the bastions and other features was known as the *trace.* As time went on the trace became more and more elaborate, other projections were added under the names of horn-works, ravelins, demi-lunes and simple and double *tenaille.* Ultimately the bastions were detached entirely and formed a ring of little forts around the main one. When the Germans attacked Belgium in 1914 they faced just such ring forts.

Aside from the self-protecting trace, the other feature of a fortification was the system of vertical

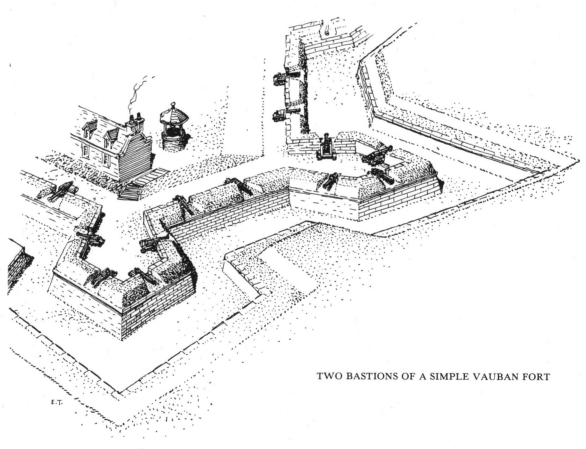

TWO BASTIONS OF A SIMPLE VAUBAN FORT

98

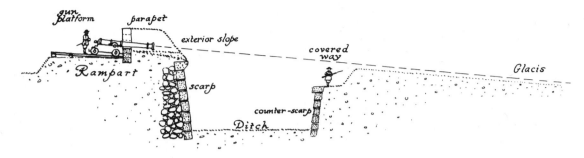

hazards placed around it to discourage the enemy and to stop cannon balls. Just as the Normans had done five hundred years before, Vauban surrounded his "bailey" with a wide ditch about twelve feet deep and corresponding in outline to the trace of the fort. The sides of the ditch were stone or brick walls, not quite vertical. The inner wall was called the *scarp*, the outer one the *counter-scarp*.

Inside the ditch the fort was surrounded by the *rampart*, which included the raised earth platform on which the guns stood, and the eight-foot-high stone *parapet* which was raised in front of the guns and slotted with *embrasures* for them to shoot through. The parapet stood well back from the edge of the ditch, and earth was banked between it and the top of the scarp to contain any metal which was thrown in that direction.

At the top of the counter-scarp on the far side of the ditch a fairly wide, level path was made where troops could be assembled for sorties against the enemy. This path was called the *covered way*. It was covered by another bank of earth high enough to keep the enemy from seeing what was going on behind it. From the top of this bank the *glacis*, or approach to the fort, sloped gradually away on all sides. The angle of the glacis slope was such that the fort's guns could rake any part of it with direct fire.

Such forts as these were built all during the eighteenth and nineteenth centuries. There are plenty of them to be seen right here in the United States. Fort McHenry of *Star Spangled Banner* fame is an almost perfect example of a simple fort built on these principles.

As soon as Vauban had invented his fortifica-

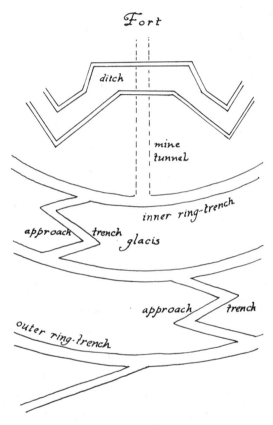

DIAGRAM OF VAUBAN'S SIEGE SYSTEM

tions, he set about finding a way to overcome them. Some say he was even better at this than at building them. The first time his system of siege was tried at Maestricht the place fell in thirteen days. The fact is that forts always fell eventually if the attack was a determined one, but sometimes they fell at such cost to the attacker that he beat himself taking them.

Vauban's system for capturing was to dig trenches into the glacis in concentric rings around

99

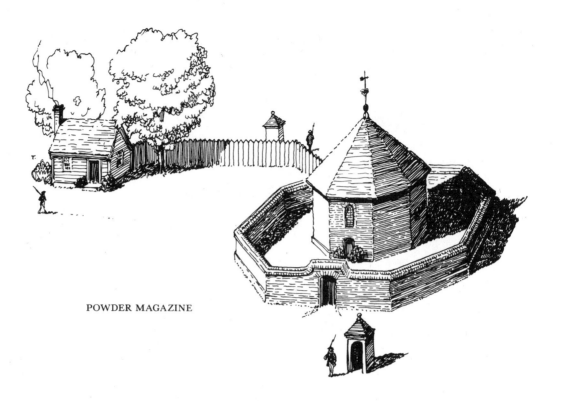

POWDER MAGAZINE

the fort, working at night and moving up from ring to ring through zigzag approach trenches. The zigs and zags were so the fort couldn't get a straight shot down a trench. The same idea was used for the same purpose in World War I.

Vauban laid down a bombardment from each ring of trenches, and from the inner ring he started mining operations which eventually passed under the whole ditch and breached the fort itself. Troops were brought in through the tunnel, and though there was still fighting to do on the inside, the fort, as a fort, no longer counted.

Every fort had its magazine. Black powder damps easily and won't shoot when it is damp, and so, when magazines were built to store the stuff, the first consideration was that they be dry. They also had to be fire resistant and strong enough to discourage invasion. So they were usually of brick or stone and were built with wooden floors raised at least two feet above ground level. The space under the floor was filled with stone chips, to allow air to circulate and yet to fill the space and guard against anyone getting in that way.

Containers of charcoal and of chloride of lime were scattered about among the powder kegs to absorb moisture from the air. Ventilation was important of course, so it was necessary to pierce the walls but the little vent holes were carefully screened to keep an enemy from tossing in fire or sending in a small animal with fire tied to its tail. To avoid the possibility of sparks, powder kegs were hooped with copper and held together with copper nails. The kegs in storage were turned over frequently to keep the heavy saltpeter from settling to the bottom and spoiling the mix.

The capture by the Colonists of the powder and muskets in the magazine at Williamsburg, Virginia, was an important early step in the American Revolution. This magazine has been carefully rebuilt exactly as it was.

100

THE GERMAN RIFLE, FORERUNNER OF THE KENTUCKY RIFLE

## The Kentucky Flintlock Rifle (1727—1820)

Little has been said thus far about American weapons, since other than those used by the Indians, all the earliest weapons in America were brought from Europe. The Spanish colonists and the first English colonists used matchlock arquebuses and muskets. At Jamestown they mounted a few cannon which duly scared the simple fishing Indians of the locality out of their moccasins.

As time went on settlers from Continental Europe began to arrive in the Colonies and they brought their own weapons with them, wheel locks, snaphances, and finally flintlocks. Most of these guns were smoothbore muskets, but not all. The Germans had been using rifles for more than a hundred years, and when they came to Pennsylvania their rifles came with them. Also, some of these settlers were trained gunsmiths.

The heavy German rifles had spiral grooves cut into the insides of their barrels. They fired a lead ball which fitted the bore so tightly that, to be loaded, it had to be driven down the barrel with a ramrod and a mallet! But once in, the ball was threaded to the grooves. It had to follow the spiral when the explosion drove it through the bore, so it came out spinning. That spin kept it on the track to its target, and it didn't bounce around in the air as a musket ball did.

Life on the Pennsylvania frontier wasn't easy. There was game in plenty but it had to be shot to be eaten. In order to sustain life it was also necessary to outshoot an occasional Indian. Every man and boy was familiar with guns and could use them. Naturally the German rifles were examined, tried and discussed. They were approved for their obvious merits and criticized for their weight, for the difficulty of loading them and for the amount of scarce powder and ball it took to shoot them.

There was something in the air of Colonial America which stimulated ingenuity. Inevitably some Leatherstocking had one of the German gunsmiths at Hickory Town (Lancaster) make him a rifle according to his own ideas. That could have been about 1720. By 1727 that first rifle and others which followed it had been tested, and each succeeding model improved until a gun had developed which had everything that anybody then hoped for. This was the "Kentucky" rifle. There's no use complaining that it was really the "Pennsylvania" rifle; it's been named and the name will stick. It was carried into "Kaintuck" when the frontier moved west, and many good guns of the type were made there, but to the end, most of them and the best of them were made in Pennsylvania.

This is the rifle that settled America and had no small part in winning for Americans their cher-

FRONTIERSMAN WITH HIS RIFLE

RIGHT SIDE

LEFT SIDE

ished freedom. It's worth a close look. Almost always it was long, five feet or more, six feet wasn't too unusual, but compared to the German rifle it was very light. It balanced beautifully on the hand. The outside of the barrel was usually "browned." The stock was maple, cherry or walnut, and the wood ran out almost to the muzzle. The stock was made very dark by rubbing it with soot and oil. The butt where the rifle rests against the shoulder and is likely to be set on the ground was protected with a metal butt plate. Usually this was brass as were the other metal "mountings" or inlays in the wood. There was some form of star set in on the left side just where the marksman's cheek rested against the stock and on the same side a screw plate which was part of the lock. Directly opposite this, on the right side, was the lock plate on which the visible mechanism of the lock was mounted; also on the right side, near the butt, was the ornamental brass cover of the patch box. The box itself was hollowed out of the wood.

In addition to being longer than the German rifle, the Kentucky had a much smaller bore, generally .44 or .45 caliber against as much as .75 for the German guns. Less hard-to-come-by lead was needed for bullets and powder was saved. Another difference was the way the bullet fitted the bore. When it was necessary to load fast, to deal with a savage who was potting at you, there was no time to swedge a ball all the way down the barrel with a mallet. So the bullet was made three hun-

dredths of an inch smaller than the bore, small enough to roll down the barrel. To give it a grip on the rifling and hence the vital spin that made a rifle better than a musket, the ball was loaded with a greased leather patch wrapped around it; a round patch about the size of a fifty-cent piece. The patch supply and a lump of grease were carried in the patch box in the stock.

LOADING A PATCHED BULLET

The patch was placed on the gun muzzle, a bullet placed on the patch and the two of them were shoved home with a ramrod in about a fourth of the time it took to load a German rifle. The loose patched ball proved to give more accuracy than the tight, naked one. In fact, in the hands of an expert and within its distance a good Kentucky is as accurate as the best of modern rifles. Its best distance is about a hundred yards, though fine shots have been made at more than twice that. No other gun of its time could hope to hit anything much smaller than an elephant from a hundred yards away.

Until after 1820 all Kentucky rifles were flintlocks. There was no special difference between their locks and the ordinary flintlock of the kind shown on page 93. They were excellent locks, but what is remarkable about them is that they were made in the wilderness: forged, hammered and filed into shape, often from iron mined on the spot and smelted with charcoal.

The whole gun was made by hand, sometimes, but by no means always, with the help of power from a water wheel. The barrel was forged by wrapping a strip of hot iron spirally around a rod and heating and hammering until it was welded into a tough tube, octagonal outside and roughly cylindrical inside. In order to get them off the

mandrel, barrels had to be made in two short lengths and welded together afterwards.

The rough inside was bored by hand because the transmission gear for the water power wasn't steady enough. Even so, the bore usually came out a little crooked and was checked with a taut bowstring and tapped gently at the high spots to straighten it. The flats on the outside were ground smooth against a revolving stone and the barrel was ready to have the rifling grooves cut into it.

Repeated cuts were made on each groove, one after another, until the cutter would bite no deeper; then the depth of the cut was increased by slipping successive pieces of paper under the tool. The saw-toothed cutter was set in a hickory stick to avoid scratching the bore. It rotated one turn in four feet as it was pulled through. The rotation was accomplished by attaching the cutter rod to a spirally grooved wooden cylinder which slid through a fixed block on which were lugs to fit each groove. When all the rifling was cut, the inside of the bore was polished with emery powder on a lead plug.

The wooden stock was roughed out with a broadax and finished with drawknife and plane. Just below the barrel the stock was drilled to carry the hickory ramrod. Every gun was provided with a mold for making bullets to fit it. In addition to

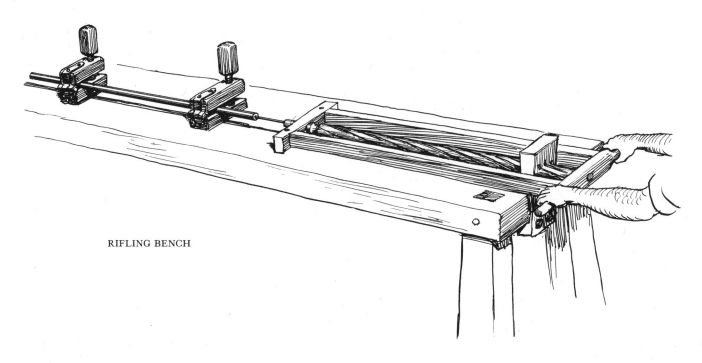

RIFLING BENCH

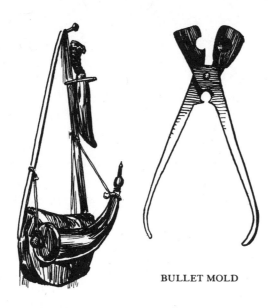

BULLET MOLD

POUCH AND POWDER HORN

his powder horn, the hunter carried a pouch in which he kept the mold, his stock of bullets, his priming pick (to clear the touchhole), a wad of crude flax (for swabbing the bore) and a twist of tobacco.

The men who shot these rifles spoke of them as "she," and out of great affection gave them female names. Those men could really shoot. It is recorded that there wasn't one man of the hundred and thirty in Michael Cresap's Rifle Company who couldn't put nineteen bullets out of twenty within an inch of a nail-head target, (presumably at sixty yards, the usual distance for show-off shooting). This was at the beginning of the Revolution and these same men, when they joined Washington at Boston, threw panic into the red-coat camp by puncturing Tory officers from distances which were supposed to be beyond gun-shot.

There was great hue and cry over this in England because it was the sort of thing that "wasn't done." Officers were supposed to be immune from sniping. To combat the American riflemen the British hired Hessians who, being Germans, were equipped with rifles. King George was badly stuck, however. The Landgrave of Hesse took his money and sent him men with rifles, but they were recruited from the plow and many of them had never handled a gun in their lives.

The British could never match the American shooting. In the War of 1812, at the battle of New Orleans, less than four thousand riflemen under Andrew Jackson took cover behind cotton bales and stood off ten thousand Englishmen. The Americans suffered 21 casualties, the English 3,336! But American marksmanship began to decline from then on because it was no longer needed in daily life, except in the far West. In World War I, American infantrymen fired seven thousand rounds of ammunition for every casualty they inflicted on the Germans. This, of course, was partly due to the kind of dug-in war it was. What the Yankees of 1776 lacked was discipline and concentrated firepower, and they couldn't win their war until they learned these things from the Baron von Steuben. In the early part of the struggle they were soundly trounced every time they came up against trained British regulars in open country.

Enthusiasts still shoot at targets with Kentucky rifles, and some of these rifles have remained in practical hunting use at least until very recently. A fine rifle in beautiful condition with all its accessories was obtained lately from an elderly mountain gentleman who had inherited it, shot it all his life and was willing to abandon it now only because his son had brought him a Mauser as part of the spoils of war.

Not all American Revolutionary soldiers carried rifles. At the outbreak of the war the Colonists were able to seize many British muskets; all those in the Williamsburg Magazine, for instance, and others elsewhere. The "neutral" King of France was good enough to send over two shiploads of French Charleville flintlock muskets, to the total number of 23,000. This addition gave the Continentals all they needed.

After the war when the Army began manufacturing its own guns at Springfield Arsenal, it at first produced exact copies of the Charleville. This was in 1795, in preparation for a war with France that happily didn't come off. In 1800 the first army flintlock rifles were made at the Harpers Ferry Arsenal. Instead of copying the fine native product, the generals hit upon a hybrid, based on

THE HALL RIFLE WITH ITS BREECHBLOCK OPEN
FOR LOADING

a German rifle. Being a hybrid it naturally kicked like a mule. Also, its barrel was too short, which made it inaccurate.

John Hall of Yarmouth, Maine, invented a *breechloading* flintlock rifle which was adopted by the army in 1819—this in spite of the fact that it leaked gas badly and, as a result, had less punch than the gun it replaced. The Hall had a hinged breechblock arranged so that its front end could be raised above the level of the top of the barrel. A chamber was thus exposed which could be quickly loaded and the block snapped down again to line up with the bore. Powder for these guns was issued in thin paper cartridges which could be emptied into the chamber. The paper was discarded. Cigarettes are supposed to have been invented by a Turkish soldier who wrapped tobacco in one of these papers when his pipe was "shot out from under him."

In the eighteenth century everyone who was or pretended to be a gentleman wore a sword. Sometimes they fought duels with them, but mostly they just wore them; pistols had become more fashionable for dueling. The day of the great swordsmen was over; even in the armies swords were used mostly for waving, though sailors put their cutlasses to practical use on frequent occasions.

The eighteenth-century dress or court sword was straight but considerably shorter than a rapier, and it was worn at a less dashing but also a less hazardous angle. Instead of the sweeping, fancy hilt of the rapier, the court sword retained only a simple ring. The quillons were still to be found, but one was reduced to a short spur and the other was elongated and bent upward over the swordsman's hand and joined with the grip at the pommel.

The seagoing cutlass, which was every sailor's

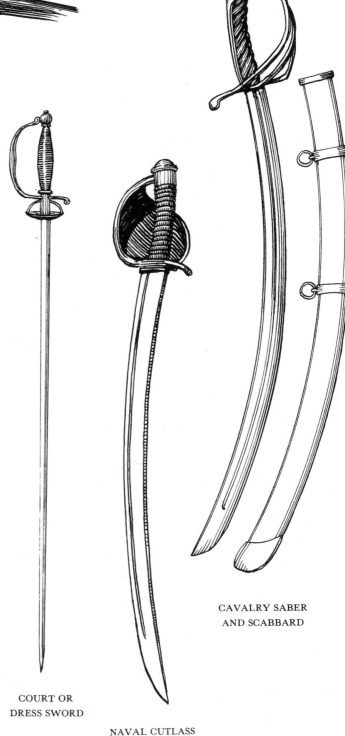

COURT OR
DRESS SWORD

NAVAL CUTLASS

CAVALRY SABER
AND SCABBARD

weapon for a boarding party, was a rather short slashing sword with a wide, curved blade, sharp on one edge. Its guard was a wide metal shell protecting the whole sword hand. Naval officers as a rule didn't use cutlasses, but carried the regular dress sword and pistols.

One other sword saw considerable use in this century—the cavalry saber. Like the cutlass, it was carried by the rank and file, but in this case by officers too. Also like the cutlass, it had a curved slashing blade which was sharp, but its blade was much longer than the cutlass and not so wide. The hilt was shaped much like that of a dress sword but heavier. Cavalry was given to wild charges, usually against infantry. During these they waved their sabers and slashed to right and left as they rode through and over the ranks. Mounted soldiers also carried short flintlock musketoons in a saddle boot and eighteen-inch *horse pistols* which fitted into saddle holsters.

Flintlock pistols came in many sizes. The barrel of an ordinary pistol was usually from seven to ten inches long. There were army and navy pistols, beautiful dueling pistols which came as matched pairs in handsome cases, and common pistols for casual social use.

Dueling pistols had sights for use in practice. In an actual duel the combatants were not supposed to aim but to raise the gun and fire on some kind of signal. In addition to Alexander Hamilton who was shot by Aaron Burr, many a good and valuable American was snuffed out in these ridiculous potting matches. Even men who fully realized the absurdity of the code hadn't the courage to refuse a challenge and face the accusation of cowardice.

## Eighteenth-Century Artillery (1700—1800)

Frederick the Great took up field artillery where Gustavus Adolphus had left it and worked out new and successful tactics for using it. He *had* to develop artillery because he'd lost so much infantry that there wasn't any more to lose. Though Frederick's army drilled in the rigid eighteenth-century way, doing everything—well, nearly everything—on command, he nevertheless thoroughly understood the value of mobility.

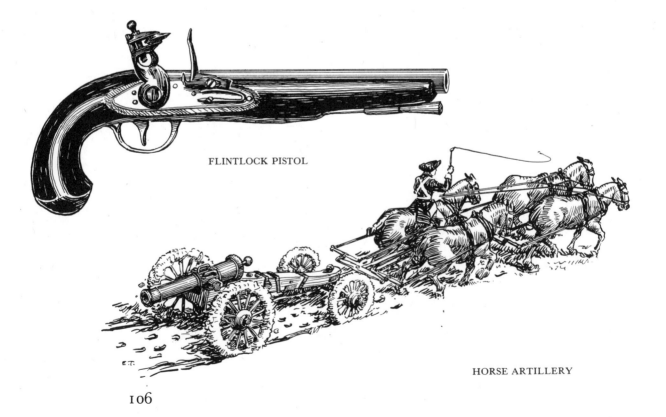

FLINTLOCK PISTOL

HORSE ARTILLERY

In most armies of his time, cannon were hauled to the battlefield with horses owned by civilian contractors, who took their nags out of harm's way before the show started. The guns stayed where they were left until the fight was over. Frederick operated differently. He set up horse artillery using army horses in charge of soldiers and trained to move almost as fast as cavalry. All his guns were light three-pounders and six-pounders. For moving, the trails of these guns were set on little carts called *limbers* to which the horses were hitched.

At that time battles were fought by ranking two armies in wide, shallow formation, face to face on an open field, and letting them shoot it out. Frederick had his artillery attack first and then brought up the infantry to charge past the cannon after the opposition had been softened up. The horse artillery would dash ahead until they were about thirteen hundred feet from the enemy. Then they'd dismount and start lobbing cannon balls at the opposing lines. Between shots the men put their shoulders to the guns and moved them forward a little. When they were close enough to make it effective, they'd change from solid shot to grape.

Grapeshot scattered like canister but it had more authority and it carried further. A charge of grape consisted of fifty or sixty iron balls, each about an inch in diameter. They were bunched around a wooden rod which had its lower end fixed into the center of a wooden disk. The whole thing was held together by a cotton bag netted with twine, and it was loaded into the gun that

**GRAPE SHOT**
*Uncovered*          *Bagged*

way. When the gun fired, the bag burned away and the balls were sprayed out from the muzzle.

A Frenchman named de Gribeauval fought against Frederick's army and came home with some new ideas. He first trained the French field artillery along the lines of Frederick's and then went on to set up specialists with special guns for siege work, for garrison (fortress) defense and for coast defense. For the last two purposes he invented the *barbette* carriage which allowed a cannon to shoot *over* a parapet instead of through an embrasure; the back of it had wheels running on a semicircular track to allow the gun to be *traversed* from side to side.

All these eighteenth-century cannon were smoothbore muzzle-loaders made of cast iron or bronze. Early in the century Benjamin Robins predicted that cannon would someday be rifled; but lead cannon balls of large size weren't practical and rifling grooves could get no grip on iron ones, so that had to wait. By the way, Robins was

BARBETTE CARRIAGE

LADLE

the first to prove that air currents affected the flight of a cannon ball.

At sea, guns were mounted on *truck carriages*. These were heavy timber frames riding on four little wheels (they were the trucks). The gun was trundled up by tackle to its porthole in the ship's side, fired and was rolled back by its own recoil until it was stopped by heavy breeching ropes attached to the back of the gun itself and to the ship's side.

The men handling a muzzle-loading cannon needed a variety of tools and equipment, some of which was used every time a shot was fired. An ordinary gun crew for a light fieldpiece would be seven men, one of whom was the gun captain. Sometimes he was simply called "the gunner." He was the expert and was likely to do the aiming and the actual firing, as well as acting as commander of the crew.

First powder had to be put into the gun. In the early part of the eighteenth century and at all times before that, this was done with a *ladle*. A ladle was a cylindrical scoop on a long handle. Level full, it held exactly the right charge of powder and it was of exactly the right diameter to fit into the bore of the gun. The ladle was charged with powder from a box or sack and pushed carefully into the bore as far back as it would go; then it was turned upside down and pulled out, leaving the powder in the gun. In the

second half of the century powder was frequently packaged in wool bags called cartridges and was put into the gun, bag and all. A similar system is still used for firing very large guns, though of course the bag is now loaded into the breech. Even after cartridges were introduced, the ladle was still useful when it became necessary to unload a gun without firing it. The metal of which the ladle was made was exactly as thick as the *windage,* that is, the space between the shot and the bore, so the ladle could be slipped under a loaded cannon ball to bring it out.

There was also the priming. This was finer grained and was brought to the gun loose in a *passing box.* The gunner put some into the vent (touchhole), and the *powder monkey* who had charge of the box took it back out of the way. Both he and the gunner were careful not to spill any because it could be annoying if it caught fire in the wrong spot. The gunner wore a leather thumbstall which he pressed down on the vent as soon as possible after the gun fired, to kill sparks. He also carried a *pick* or priming wire to clear the vent. After cartridges came in, he used the pick to push down through the vent and punch a hole in the woolen bag, so fire could be sure of reaching the charge.

If a cartridge was used, the *rammer* served to push it down the bore, and even with loose powder it was used for shoving the stuff well back,

108

and for pushing in the wad and the shot. The wad was put between powder and shot and was usually rags, cotton waste, or even straw. The shot was carefully inspected and cleaned before it was put into the muzzle because a little dirt could ruin the bore or might even cause the gun to blow up. The rammer was simply a short wooden plug which fitted the bore and was fixed on the end of a long handle. The handle was ordinarily scored with marks to show when each part of the load was properly seated.

The actual firing was done by applying the old slow-match to the vent hole. The match was always in charge of the gun captain who carried it on a stick called a *linstock,* so that he could stand clear when he touched off the priming. The linstock had a ring or a clamp at one end to hold the match and a metal spike at the other for standing it up.

After the gun had fired, the bore was swabbed out with a *sponge* which of course was *not* a sponge. It was another long-handled wooden plug but slightly thinner and considerably longer than the rammer, and it was covered with sheepskin, woolly-side out. Sometimes on the end of the sponge handle, sometimes on a handle of its own, the *wormer* was an iron double screw, used as needed, to remove bits of old wad and unburned scraps of wool powder bag.

A bucket of water stood near the gun, and the sponge was dipped into it before swabbing so that it would be sure to kill any sparks which might be alive in the bore after firing. A new charge of powder rammed in on a hot spark could be a real nuisance to the man with the rammer. Firing made the barrel hot enough to evaporate the moisture from the sponge before the next charge was put in, the barrel actually heated so much that it wasn't safe to load at all after forty rounds. The gun then had to be cooled off for an hour. Even so, light guns sometimes fired a hundred rounds a day; thirty was good for a heavy piece.

There was another article in constant use—the *handspike.* This was a crowbar or pinch bar of wood, shod with iron. Ashore it was used to heave the trail around for aiming from side to side. At sea a handspike was used to gain the same end but the whole back of the gun carriage had to be lifted. A ship's deck around a gunport was always

chewed up from handspikes. In the hands of a husky sailor a handspike was no mean weapon in itself.

These were the main tools. There were still others: a *scraper* made as two half-disks for getting rust and hardened soot out of the bore; a *cat* of springy wires for hunting defects in the bore; and a *searcher* which could take a wax impression of a defect to determine how bad it was. A cat had two handles, one of which operated a sliding ring for contracting the "feeler" wires to free them if they stuck in the bore. At the end of firing, a lead plug went into the vent hole and a kind of pot lid called a *tompion* was fitted into the muzzle, both of them to keep out dampness.

Even before the American Colonists were organized to fight the Revolution, they began to round up all the cannon they could lay hands on and all the powder and shot they could find. As a result Washington's army used guns of thirteen different sizes. Not enough guns were "liberated" for fighting a war, so little iron foundries, located

RAMMER

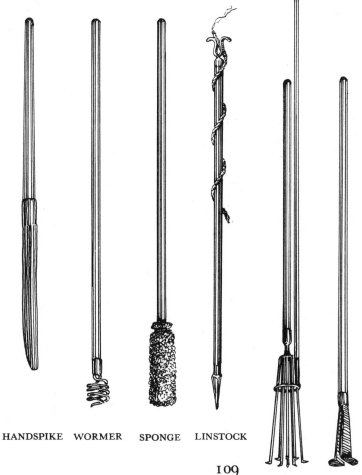

HANDSPIKE   WORMER   SPONGE   LINSTOCK

109

"CAT"   SCRAPER

near falling water where power could be had, began to cast cannon for the Congress. You may still come upon a "Gun Road" here and there in the East, and always it leads uphill from some rushing stream.

No cannon were imported during the Revolutionary War. The fieldpieces ran from three-pounders to twenty-four-pounders, the smaller ones being bronze. They were moved to the field by horses or oxen in charge of civilian drivers and then hauled around the field by the gun crews pulling on drag-ropes. We hadn't quite caught up to Frederick at that time.

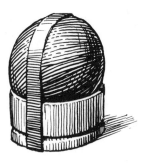

SEMI-FIXED AMMUNITION—AN IRON BALL BANDED TO A WOODEN SHOE

By 1790, though it was still loaded into the muzzle, ammunition began to be prepared in advance for the guns. Some was what is called *semi-fixed;* it was simply a round shot strapped to a hollowed wooden disk which served as a wad. *Fixed* ammunition also appeared. This had not only shot and *sabot* as the disk was called, but also bagged powder, all put together as a unit.

## Percussion (1800—1850)

The wheel lock was better than the matchlock, and the flintlock was simpler and cheaper than the wheel lock, but even the flintlock was far from perfect. It often failed to go off. In 1834 the British army got around to testing their old Brown Bess against a new percussion musket. Six thousand rounds were fired from each gun. The new gun

failed to shoot six times when its trigger was pulled; Bess missed fire nearly *one thousand times.*

The percussion idea was hit upon by a Scottish clergyman, the Reverend Alexander Forsyth, who was experimenting to find a way to apply the newly discovered fulminates to the priming of guns. When he found that these substances would explode more readily if they were struck a sharp blow than if they were ignited by flame, he got his idea, and in 1807 he patented the first gunlock which made use of the principle. Though his lock has long since been discarded, practically all modern guns big and little, are fired by the percussion principle.

Dominie Forsyth's lock was simple and clever. A round plug was set into the side of a gun barrel. In the upper side of the plug there was a small recess which served as a flash pan. A hole was drilled straight down in the center of the pan to connect with the end of a horizontal hole which led to the gunpowder in the bore.

The plug projected about an inch from the side of the barrel, passing clear through a metal gadget which came to be called a "scent bottle" and which could be rotated by hand on the plug. The lower end of the scent bottle was a storage container for fulminate, and held enough for priming twenty shots or so. To prime, the bottle was simply turned upside down for a moment and fulminate was dropped from the storage recess into the pan. Turning the gadget right side up again put the stock of priming out of reach of sparks (there were some accidents however) and brought the firing pin into position directly over the flash pan.

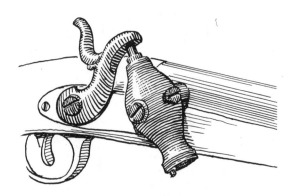

FORSYTH "SCENT BOTTLE" PERCUSSION LOCK

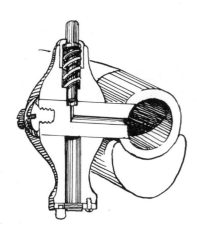

THE INSIDE OF THE SCENT BOTTLE

A coil spring held the pin up a fraction of an inch off the priming, and all that was needed to fire was a light blow on the projecting upper end of the pin. This was delivered by a hammer, which landed on the firing pin when the trigger was pulled. A true hammer this time, the first one.

Forsyth set up in business in London with James Watt, of steam engine fame, as a partner. Of course once they saw the idea, every gunsmith wanted to get into the act and poor Forsyth spent his profits and the rest of his life suing people who infringed upon his patents.

Many legitimate variations of percussion locks were invented. Some used a little tube of fulminate, some a pellet which could be placed in a flash pan above a hole leading to the chamber and simply struck by the hammer itself. The best one was invented by an English artist named Joshua Shaw who was living in Philadelphia. Wouldn't you think it would be gunsmiths, sportsmen or soldiers, and not preachers and artists, who made improvements in guns?

Shaw originally made the thing for his own use. His gun had no flash pan at all; instead, it had a small steel nipple stuck upwards from its barrel. A little hole passed clear through the nipple and into the powder chamber. First Shaw had a little steel cap which fitted over the nipple and which he primed with a fulminate pellet for every shot. Later he tried pewter caps, loaded ahead of time and thrown away after firing. Finally, in 1816 he substituted copper and perfected the cap lock which put all others out of business and was used unchanged for fifty years on both sides of the Atlantic. It was easy to convert a flintlock to percussion and thousands of them were changed over.

Though the gas-leaking problem wasn't yet licked, it was easier to make a percussion breechloader than it had been to make one with a flintlock. In 1831 the first breechloading shotguns appeared. Speaking very strictly, a shotgun is a smoothbore musket. Shot was often fired from muskets and a shotgun *can* fire a solid ball, but a shotgun usually is charged with a thimbleful of lead pellets loaded in a paper "shell." The pellets may be as fine as sand or as big as peas. Early shells were fired by letting the hammer hit a little pin which was incorporated in them and was connected with a fulminate cap on the inside. The explosion forced the paper shell tight against the walls of the bore and formed the first effective gas seal for a breechloader. Gilbert Smith, an American, invented a carbine in 1835 which "broke" just behind the chamber on a hinge and was locked shut after loading by a bar controlled by an extra trigger. It leaked gas badly and wasn't too successful as a carbine but the idea proved to be a natural for shotguns and has been used ever since.

The Prussian needle gun was invented in 1838

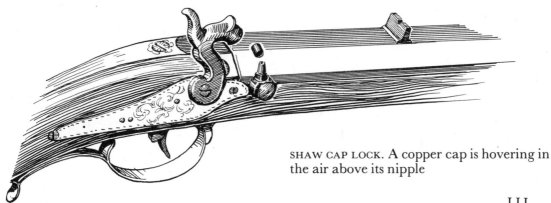

SHAW CAP LOCK. A copper cap is hovering in the air above its nipple

111

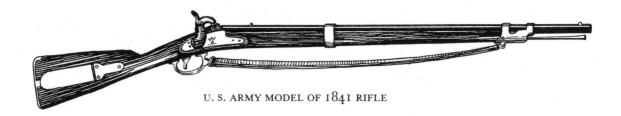

U. S. ARMY MODEL OF 1841 RIFLE

and was a breechloading rifle. It was loaded with a paper cartridge and its peculiarity, which gave it its name, was that the hammer actually drove a long needle all the way through the gunpowder and into a fulminate cap which was placed between powder and bullet. This was supposed to cause a faster and more complete burning of the charge.

The U.S. Army, when it adopted percussion in 1842, stuck to muzzle-loading and to Mr. Shaw's cap lock; they paid him $18,000 for the use of it. The rifle they produced was for many years the best military arm in the world. It was very accurate and its bullet would penetrate eight inches of pine at a hundred yards.

Many of these "Model of 1841" rifles were made in army arsenals, but many more were made by civilian gunsmiths under contracts. One of the contractors was Eli Whitney, who invented the cotton gin but also thought up the much more important idea of making gun parts so accurate that any piece would fit perfectly on any gun of the same model. This is commonplace now of course with everything that is made on an assembly line, but when Eli scrambled the parts of two weapons and reassembled two good working rifles from the mix, he startled the world.

Patching every rifle bullet was at best a nuisance and even with the spin it acquired from the rifling,

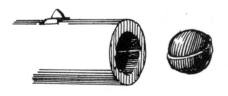

A BELTED BALL AND THE BORE IT FITTED

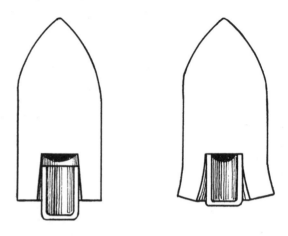

DIAGRAMS OF MINIE BALL UNEXPANDED AND EXPANDED

a spherical ball didn't always fly true. Then too, a molded ball usually cooled with a dimple in one side which tended to make it lope through the air. Lots of ideas were tried: balls with raised belts cast on them to fit a barrel which had only two grooves; conical bullets with four lugs on them to fit into four grooves; even *oval* bullets which fitted a twisting oval bore.

The French made the first real improvement with the Minié ball which wasn't a ball at all. It was a "bullet-shaped" bullet, with a pointed nose and a flat base. Not the first of that shape, but the first which did what it was supposed to do. It was its efficiency, not its shape, that put the Minié ball on the map. A fairly deep hollow was cast into its base, and when the bullet was loaded, a little iron thimble was put into the mouth of the hollow. The thimble didn't reach the bottom of the dent, so when the powder expanded the thimble started forward a fraction of an instant before the bullet moved. This jammed the thimble all the way to the bottom of the depression and forced the lead walls outward into the rifling grooves. The entire expanded base was exposed to the pressure of the gases and no part of them could leak past the bullet; it got all the push there was. A Minié ball of the same weight would carry further than the old ball, with much greater accuracy. The Minié had faults of course. Now and then an iron thimble would be driven clear through a bullet, leaving a nasty lead ring stuck tight in the barrel.

Many experimental breechloaders appeared. There was an English one in which the whole barrel moved forward to allow it to be loaded. They all leaked gas at the breech, including a famous American one, the Sharps carbine. Its breech block was moved up and down by the action of a hinged trigger guard. The edge of the breechblock cut the end off a paper cartridge as it closed and, in doing so, always spilled a little powder. There were fireworks when the Sharps leaked gas! Nevertheless it was accurate and very quick to load and fire. Priming was done by fulminate pellets made up in a paper tape, like caps for a cap pistol. This was the invention of a Washington dentist named Maynard (not a gunsmith you notice). When John Brown tried to capture the Harpers Ferry Arsenal, his men were armed with Sharps guns. The British army adopted them and they were in use by Union snipers in the Civil War. It is from them that the word "sharpshooter" comes.

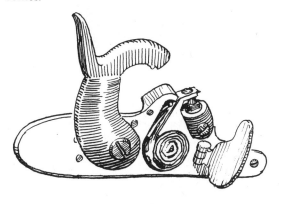

THE MAYNARD TAPE PRIMER

Cap locks made pistols much handier weapons, and there was a notable increase in the use of them in the United States, not only on the Western frontier but throughout the East. There was nothing unusual then in an ordinary citizen going about with a gun in his pocket, and those in dangerous professions like gambling invariably toted a

DERRINGER

PEPPERBOX

"shootin' arn." Derringers with an inch or so of barrel were favored by the card-playing fraternity because they were easily concealed in a pocket. They delivered one shot half an inch in diameter; of limited range but, well placed, it gave a man caught with five aces a chance to get out of the room.

There were longer-barreled pistols which could be carried on horseback or on the seat of a gig, and there were single-shot military pistols in profusion. The need in an emergency for more than just one shot produced double-barreled pistols and also "pepperbox" pistols which had four, five or six barrels. These were loaded separately from their muzzles. The barrels were usually bored into a solid cylinder of metal which rotated on a long steel pin. Sometimes the barrels had to be rotated by hand, but most of them were turned by the action of pulling the trigger which also raised and dropped the hammer. This is called "double-action."

General Sam Houston had a percussion rifle which fired half a dozen shots from a chambered block of steel which moved harmonica-wise across the barrel, lining up one chamber after another.

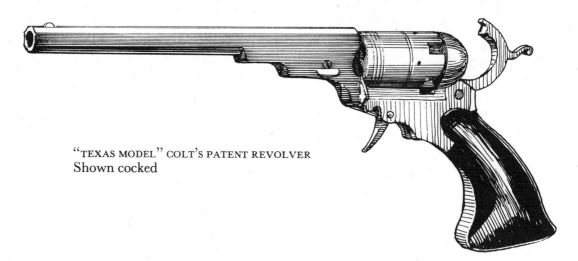

"TEXAS MODEL" COLT'S PATENT REVOLVER
Shown cocked

Rifles with revolving chambers were also made but none of them were anything more than ingenious tries.

The first repeating arm to come into general use was the famous Colt's Patent Revolver, invented in 1836 by Samuel Colt of Hartford, Connecticut. This was a single-barreled pistol with five chambers bored into a revolving drum. Each chamber was separately muzzle-loaded and had its own nipple for use with a copper cap; each in turn was moved into alignment with the barrel by the action of cocking the hammer. The first Colts had no trigger guard and the trigger itself was concealed in the stock until the piece was cocked. They were "single-action" and the hammer had to be pulled back to cock them.

Revolvers had been made before Colt's time. His accomplishment was to make good ones on a production basis. They came to be made in many sizes and models and the mechanism itself was altered for special purposes and for Service weapons. Dragoons and cavalry used Colts in the Mexican War and the Texas Rangers performed prodigious feats with them against the Comanche Indians. In one case fifteen Rangers defeated eighty Comanches, killing forty-two of them. This was considered miraculous; no one had ever heard of such firepower.

## The Rockets' Red Glare and the "Soda Bottle" (1800–1850)

Most of the cannon on the American side in the War of 1812 were of small caliber and limited

range. On the famous night when *The Star Spangled Banner* was written, the British ships shelled Fort McHenry from two miles down the harbor. "The rockets' red glare, the bombs bursting in air" which showed Key the flag were all fired *at* the fort. The defenders, knowing that the British fleet was hopelessly out of range of their guns, very sensibly holed up underground and saved their ammunition. The one attempt by the British to land and take the place from the rear was driven off.

EIGHTEENTH-CENTURY BOMB WITH RINGS

The "bums," as "bombs" was pronounced then, were the same kind of hollow, iron shell filled with gunpowder that had given bombardiers trouble in the seventeenth century. Though Mr. Key used the older name, these things had come by his time to be called "shells." In the eighteenth century they had been made with two little rings on either side of the fuse hole for easy carrying; these were now replaced by two recesses which could be gripped by the points of carrying tongs, very much like ice tongs. Only the fuse had really improved. It was now a rather long, tapered, wooden plug drilled all the way through and filled with caked powder. The bombardier padded the top of the fuse with

114

HANDLING A BOMB WITH TONGS

small war boats known as *bomb ketches*. Their peculiarity as sailing vessels was that their forward decks were clear and their forestays, which braced the masts, were made of chain to reduce fire hazard.

The British picked up the idea for their military rocket in India late in the eighteenth century and produced iron-headed rockets that would carry an explosive black powder charge more than two miles. A thirty-two-pounder had a three-and-a-half-foot head fastened to a fifteen-foot-long guiding stick which traveled with it. The rocket was launched from an inclined trough and was exploded on landing by a crude time fuse. These rockets seem to have been more spectacular than dangerous. Iron rockets were not dropped by the British until World War I, but by then they had lost their sticks and were held on course by the rotation given them by three little vanes in the jet stream. Now the rocket is back and has among other things, increased the firepower of the foot soldier for short ranges above that of the old field artillery.

CONGREVE
MILITARY ROCKET

tow, put a fuse setter above it and drove the fuse down into a hole in the bomb with a mallet. It was then "single-fired," that is, fired fuse outward and ignited by the blast of the gun.

Paper fuses were soon invented which could be inserted into a similar plug. They contained an inflammable composition whose rate of burning was determined by its mixture and indicated by the color of the paper. Thus a shell for a short-range shot could have a faster fuse than one for a long shot. Some fuses of this kind were used in this country until after 1900.

The bombs which landed on Fort McHenry were fired from mortars emplaced on the deck of

A better method of firing a cannon appeared in the early eighteen hundreds which put a final end to the smoky old slow-match, except for shooting off fireworks. The new gimmick was called a *friction primer* and it worked like an ordinary kitchen match. The body of the primer was a powder-filled copper tube which could be stuck into the vent hole of a gun. Later, because they tended to enlarge the vents, they were made to screw in. A roughened, twisted wire passed through a hole drilled across the tube near its upper end. This hole was filled around the wire with a sulphurous substance, like that of a matchhead, which would

ENGLISH MILITARY ROCKET ABOUT 1900

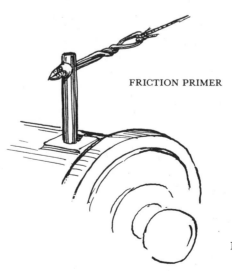

FRICTION PRIMER

THE FOREDECK OF A BOMB KETCH

CAVELLI SHOT

WHITWORTH SHOT

ignite from friction. The gunner hitched a short line called a *lanyard* to a loop in the end of the twisted wire, and to fire, simply gave it a sharp yank.

There was also a percussion primer for cannon. It had a fulminate cap in the top of a powder tube. Pulling the lanyard caused a hammer to hit the cap.

Robins's old prediction of a rifled cannon was made an actuality in 1846 by an Italian named Cavelli. His gun had only two spiral grooves and its cylindrical shot had four projecting studs, two on each side, staggered just enough to make them ride the grooves. Joseph Whitworth of England built a gun with a twisting hexagonal bore and a long shot made with six flats angled to fit. Both of these guns were breechloaders.

However, the day of the muzzle-loading cannon wasn't yet gone. John Dahlgren unveiled his "soda-bottle" shell gun in 1850. It was a smooth-bore muzzle-loader; Dahlgren also invented a rifled howitzer. The soda-bottle was made of cast iron and it achieved its outside shape as the result of a study of the varying pressures *inside* the bore. It was thick where it needed to be thick and thin where thickness didn't matter. The result looked like the soda-water bottle of the time. Dahlgren wouldn't have cared if it had looked like a turnip; he was after results. The soda bottle was mounted on a Marsilly truck carriage designed for use with a special roller-tipped handspike. Instead of the old wedge-shaped quoin for controlling elevation, there was an elevating screw threaded through the back of the gun. Otherwise the gun was handled and fired just as guns had been fifty years before.

DAHLGREN "SODA BOTTLE" ON MARSILLY CARRIAGE

## Gastight Cartridges and Smokeless Powder (1850–1900)

When the United States decided to adopt a bullet-shaped bullet for its Army rifles in 1855, a smart American mechanic licked the fault of the Minié ball with nothing more complicated than leaving out the iron thimble entirely. The gases expanded the rim of a hollow-based bullet just as well without the thimble and they never blew a hole through the lead.

Dr. Maynard, the same man who designed the tape primer, built a breechloading carbine which saw service in the Civil War. It used a metal cartridge pierced at its back end to admit fire from an outside priming cap. So there was still a gas leak through the hole. An American colonel named Berdan borrowed a French idea and succeeded in making a metal cartridge with a built-in primer. Maynard adopted it for its simplicity and safety and then discovered that with the cap plugging the hole, he had achieved the obturating or gas-sealing cartridge. The expanding gases of the explosion pressed the thin walls of the shell so

tightly against the sides of the chamber that they prevented their own escape. All guns except very large cannon are now sealed by this principle.

In addition to Maynard's carbine, other breech-loaders were used in small numbers in the War Between the States. One of them was the Sharps rifle mentioned earlier. There were even repeaters, but militarily these special guns counted for little

THE MAYNARD CARBINE AND ITS PIERCED CARTRIDGE

117

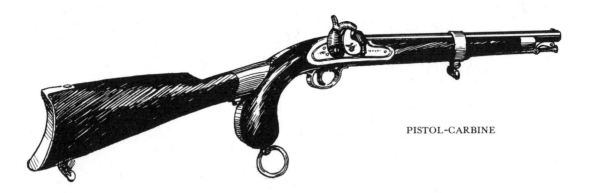

PISTOL-CARBINE

in comparison to the thousands of muzzle-loaders in the hands of the half-trained troops, many of whom had never fired a shot before. Some twenty thousand guns were left on the field at Gettysburg. They had been jammed with one load on top of another by nervous soldiers and then abandoned. Of course there was no immediate way to tell whether a muzzle-loader was charged or not, since you couldn't see into the barrel. So after the war the Army changed over to breechloading for all rifles and carbines.

One Civil War gun of small importance but of some interest was the pistol-carbine authorized by Jefferson Davis when he was United States Secretary of War. This was a muzzle-loading gun with a twelve-inch rifled barrel fired by a tape primer or by copper caps. In addition to its ordinary curved pistol grip, it had a detachable shoulder stock. This piece was issued to dragoons and to cavalry. The dragoons carried it on horseback in two parts but, dismounted, they used it almost entirely as a shoulder gun and liked it. The cavalry tried to use it as both pistol and gun, never got the hang of it either way and cordially detested the thing.

The advantages of the expanding metal cartridge were so obvious that the Army designed a breech converter to change old muzzle-loading rifles into breechloaders. Colt revolvers were also changed over by thousands, both in and out of the service.

The first breechloading revolver which was designed and built to use metal cartridges was the Model One, Smith and Wesson. It was a seven-inch .22 caliber with seven chambers and a rifled barrel. It was manufactured from 1857 to 1860. Caliber, by the way, is the diameter of the bore in hundredths of an inch. In 1869 Smith and Wesson made a longer .44-caliber gun for the Army. Afterwards Army pistols were usually .38's, but in the Philippine Insurrection of 1900 it was found that this size didn't hit hard enough to stop a berserk native in his tracks, so .45's were adopted and have been used ever since.

For some reason that's hard to understand now, the armed forces still found a use for single-shot pistols after the invention of the revolver. Reming-

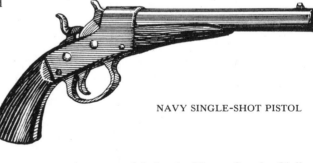

NAVY SINGLE-SHOT PISTOL

MODEL ONE,
SMITH AND WESSON REVOLVER

ton made two models for the Navy after the Civil War and one for the Army as late as 1871. From here on revolvers and cartridge rifles were made by every gunsmith; some were better than others and they differed in minor points of design.

THE KRAG-JORGENSON—U. S. MAGAZINE RIFLE,
MODEL 1892

After examining more than a hundred designs for breech mechanisms submitted by inventors in 1872, the Army vetoed all of them and settled on a design which, in the main, was a copy of the result obtained by converting the old muzzle-loaders to breechloading. This design was known as the Springfield rifle, and with occasional improvements it remained the standard arm until 1892. At its best it was a good gun.

Ten of the guns submitted to the Board of 1872 were magazine rifles, or repeaters. As soon as expanding metallic cartridges were invented, the possibility of feeding them mechanically into a barrel became obvious. The earliest design was made in 1849 but it hadn't much success until it was applied to the Henry rifle in 1860. This gun carried fifteen cartridges in a tube under the barrel. From there they were fed into the breech one at a time, by a mechanism operated by swinging the trigger guard forward and back. This system, in fact this rifle, became the basis of the famous Winchester repeater.

The Henry rifle and one other repeater, the Spence which carried its eight shots in a tube bored into its butt, were tried out in the Civil War by volunteer troops who bought their own guns. In spite of a tendency of the magazines to explode, the Army was impressed. It estimated that one soldier shooting from cover with a Henry was worth eight soldiers in the open with regular rifled muskets. The Henry was acknowledged the most effective military arm in the world. The Confeder-

ates called it "that damn' Yankee gun that can be loaded Sunday and fired all week." Still, repeaters were not adopted officially.

In 1882 the Ordnance Department invited inventors to submit designs for repeating rifles. Of the forty submitted three were thought good enough to try out under service conditions, but though one of them was a Winchester, none of them was thought to perform satisfactorily. There is more to picking a service arm than merely finding one that will work. It must be tough and it must be simple enough for the average soldier to handle and keep in working order. In 1892 the government bought some Krag rifles from Norway and started making modified copies of them. These would deliver five shots in a hurry but each with less range and less accuracy than the old Springfield's one. Improvements were made and the Model of 1896 was a somewhat better gun.

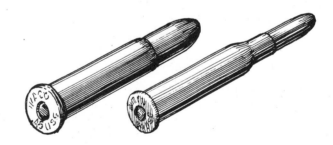

.45 CALIBER METALLIC CARTRIDGE USED IN THE
U. S. MODEL 1873 SPRINGFIELD RIFLE AND .30
CALIBER CARTRIDGE FOR KRAG RIFLE

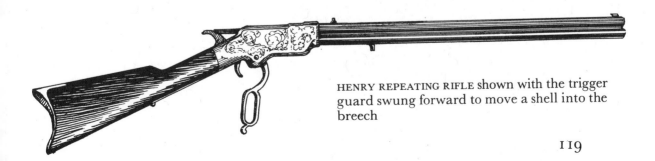

HENRY REPEATING RIFLE shown with the trigger guard swung forward to move a shell into the breech

119

For a repeater, black powder wasn't much good. It made so much smoke that after a few shots the rifleman had his head in a cloud and was likely to be unable to see his target. Its tendency to foul a barrel was bad in any case but worse for rapid fire. The invention of "smokeless powder" solved these difficulties as well as the bugaboo of powder that wouldn't shoot because of dampness; under some conditions smokeless powder could be fired even soaking wet. It was originally given the name of "gun cotton" because it was made by dissolving cotton in nitric acid. Other ways have been found to make it and some variation of it has now replaced black powder as a propelling charge for all sizes of guns. It fouls a barrel far less and it's four or five times as powerful as the old charcoal-sulphur-saltpeter mixture. It opened the way for really rapid fire and permitted the development of automatic pistols and machine guns.

An automatic pistol is really a self-loading pistol. It doesn't fire itself. Various near-successful ones were tried in the eighties and nineties, but a man named Borchardt from Connecticut really invented the type in 1893. Various features of his gun are retained in all automatics today: for instance, the metal magazine which slid into the butt, and the use of the energy of the recoil for extracting and loading. To accomplish this the barrel was allowed to slide back a limited distance after firing. This gun was manufactured in Germany, and though the original model was very heavy and badly balanced, it was by modifications of it that the wonderful Luger was developed.

## Rifled Cannon and Recoil Mechanisms (1850–1900)

In the 1850's everybody experimented with rifled cannon. Most of the iron projectiles were surrounded with lead jackets to make them seize the rifling grooves. In this country these guns were practically all muzzle-loaders, but in England and Germany some breechloaders were tried. The Americans "stuck to their guns" on muzzle-loaders for a long time; and the British, after nearly

BORCHARDT AUTOMATIC PISTOL

LIGHT PARROTT RIFLED CANNON

ten years of trial, returned to muzzle-loaders and stayed with them until a gun crew accidentally put two charges into one gun and blew away themselves and a large chunk of one of Her Majesty's ships. You can't double-load a breechloading rifle.

One of the experimental British rifles was built up by an adaptation of the old barrel-stave-and-hoop system of the bombards. Pressures in a rifled barrel where there was no windage were too great for a simple casting to take. In our own day they run above twenty-five *tons* per square inch, and all cannon have to have some kind of very special construction to stand the gaff.

Robert Parrott of the United States designed a cast-iron rifle which had heavy wrought-iron bands around its breech to hold it together at the point of maximum strain. General (then Captain) Rodman, also an American, devised big smoothbores something like the Dahlgrens, which were cast around a chilled core so that the inside surface was hardened first and then squeezed by the contraction of the outside metal. Wrapping with steel tape has been used successfully, though it makes a gun "whip" when it's fired and tends to add to the "droop" of a long barrel. Large modern guns are either shrunk over a steel liner by being expanded by heat and then allowed to cool in place, or have their bores expanded to size by terrific hydraulic pressure.

The Parrott rifles gave good accounts of themselves in the Civil War as field guns, as siege guns and as naval guns. They were made in seven sizes, from ten-pounders to three-hundred-pounders. It was a group of the larger Parrotts which hammered Fort Sumter to pieces from two miles away.

Instead of a lead jacket, the Parrott projectile

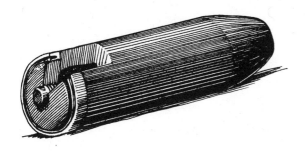

SHELL FOR A PARROTT RIFLE, cut away to show brass ring

THIRTEEN-INCH CIVIL WAR MORTAR

had a flat brass ring near its base. The upper half of this band hugged the shot, but the lower half was undercut a little so that the gas pressure could get behind it and force it outward into the rifling. Having to be slid into the barrel from the muzzle, these projectiles fitted the bore so loosely before the flange was expanded that they could slide forward if the gun was stopped suddenly when it was being "run out" to fire. If this happened it spoiled the shot. The air space it created would cushion the gas pressure so that it wouldn't expand the flange. It was even possible, if the barrel was depressed below the horizontal, for the shot to slide clear out of the bore before it was fired!

The Rodman smoothbores were used mostly for siege work and for coast defense. Some of the fifteen-inch size can still be seen in some of the old forts which have been preserved as monuments. The largest Rodman had a *twenty-inch* bore and fired a shot weighing more than a thousand pounds. Along with these big fellows in the coast forts and at sieges were some mortars with thirteen-inch bores which could toss a ball two and a half miles.

General Rodman improved the burning of black powder by molding it into "mammoth" grains three quarters of an inch thick. These burned more steadily than corned powder and gave the projectile something of a steady shove rather than a sudden boot. Before the Civil War was over smokeless powder appeared, and soon afterwards it eliminated black powder except as fuse material.

Frederick the Great and Napoleon were able to get their field guns close enough to blast the enemy infantry with case shot and grape. These didn't carry very far even from the best of guns. In the Civil War it was discovered by both sides that infantry rifles were so good that they kept the artillery too far off to use anything but solid shot. This led to the later development of explosive shells and shrapnel, both of which could be fired to the full range of a rifled piece, and it fostered the machine gun.

The Germans had pioneered in the use of steel for rifled cannon. At Sedan in 1870 their breech-loading rifles chewed up the third Napoleon's whole army. After that the use of steel became general, and the smoothbore ended its career. Muzzle-loading went out for good when the interrupted-screw breech block was perfected. Shells could then be made large enough to fit the full diameter of the bore. You could also shoot downhill without embarrassment. Shells were inserted in the back of the gun and the powder chamber was made just a little larger than the bore. Parrott's brass ring was replaced with a copper rotating band which was actually larger than the bore, but which would pass through the chamber. When the gun fired, this soft band was forced into the rifling grooves by the explosion and the grooves cut channels into the copper.

This breechblock in the illustration is of the slotted-screw or interrupted-thread kind. Gun breech and block have matching threads which are cut by six spaces so staggered that the block may be slid directly into the breech. Turning the crank will swing the block on its hinge, slide it from its "tray" into the breech and rotate it the one-twelfth turn needed to lock it. Later breech blocks of this class have stepped threads on three levels so that the meshing is continuous all the way around the breech. The inner end of the breechblock has a springy metal "pad" which seals in gas pressure.

Modern fixed ammunition came along about this time. This had its *propellant* (powder) in a brass case attached to the projectile; the case expanded like an ordinary rifle cartridge and sealed the breech. Ways were now thought up for absorbing the shock of recoil and returning the gun to firing position without moving the carriage. In other words, the gun was able to move backwards on its carriage instead of moving the whole business back with it.

One of the earliest recoil guns was the "disappearing gun" for coast defense. Many of them were emplaced along our coasts. They were big rifles mounted on counter-balanced carriages.

BREECHBLOCK FOR A BIG GUN

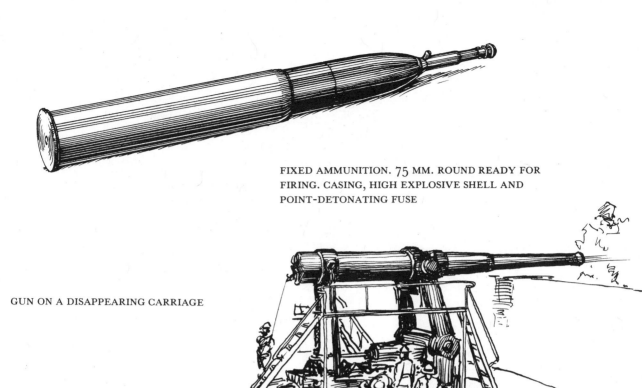

FIXED AMMUNITION. 75 MM. ROUND READY FOR FIRING. CASING, HIGH EXPLOSIVE SHELL AND POINT-DETONATING FUSE

GUN ON A DISAPPEARING CARRIAGE

They could be serviced and loaded behind a parapet and then raised by the counterweight to fire over its crest. The recoil energy which brought them back to loading position was absorbed and cushioned by the resistance of oil in big cylinders. The oil flowed through holes in pistons, and as there was means of gradually reducing the size of the holes, the resistance was increased to "set her down easy." It wasn't quite as simple as this, of course. At one time it was planned to use these as naval guns but it was never done. Their great fault was that their muzzles couldn't be elevated enough to take advantage of their full range. Modern pinpoint air bombing put an end to their usefulness.

In 1897 the French came up with their great "75" which immediately put all older field guns on the shelf and which, in modified form, was used by them and their allies including the United States, until World War II.

The "75" was named for the seventy-five millimeter diameter of its bore, just a fraction under three inches. Later ones had steel, balloon-tired wheels and split trails but the original had a solid trail and wooden wheels. At the back of the trail was a spade which could be dug into the ground to keep the gun in position. It was aided in the job by two connected brake shoes which could be dropped under the wheels to lock them.

The gun itself was supported by a steel cradle in which it was free to recoil on four rollers and there were two more rollers under the muzzle to support it in the cradle when the gun was all the way back. There were no trunnions on the gun itself. The barrel was connected to a piston which moved in a cylinder bored into the cradle. As the recoil of the gun moved the piston back, oil behind the piston was forced into a second cylinder where it compressed nitrogen gas. When the recoil was complete, this gas expanded and, reversing the whole cycle, brought the gun back into firing position. All this was fast. *Bang!*—Bump! And the gun was ready to load again.

The breechblock on a "75" was opened by

123

simply rotating it. Part of it was cut away and when this break was turned to the top it lined up with the bore, like the top of a tooth-powder can. At just the right point in turning, it struck a cam which extracted and ejected the empty cartridge. The firing pin moved with the breechblock, and there was a spring hammer operated by a lanyard. The hammer struck the firing pin and drove it forward against a percussion cap, just as a rifle is fired.

The gun was provided with an accurate telescopic sight. Turning a hand wheel would move the gun from side to side on its axle, and a couple of other hand wheels raised or lowered the barrel to the needed angle of fire. At nineteen degrees above the horizontal, which was as high as the elevating mechanism would take it, the gun could shoot more than a mile and a half; but if you put its trail down in a ditch and thus elevated it to forty-four degrees, it would carry a thirteen-and-a-half pound HE (high explosive) shell nearly two and a half miles. Gas shells and shrapnel were later fired from 75's.

In its maturity in the U.S. Army the "75" was put on pneumatic tires for high-speed travel, and the trail was split into two halves which could be joined for moving but which were separated when the gun was set up to fire. Separating them allowed the gun to reach its top range. The standard "75" finally faded because it couldn't knock out im-

proved German tanks in World War II, but it has modern descendants with tricks the old field gun never dreamed of.

## Quick-Firing and Machine Guns (1850—1900)

The advantage which infantry had attained over artillery gave trouble in Europe even before it appeared in this country, Napoleon III practically ordered his man Montigny to do something about it. Amid great hush-hush and broad hints of invincible secret weapons, Montigny brought forth the *mitrailleuse*.

It was mounted on a gun carriage and was pulled by four horses. At a glance it seemed to be a field gun but what looked like its barrel was really thirty-seven barrels in a bunch. A magazine containing a positioned load for each barrel slipped into a space at the breech end. Each barrel had its own firing pin which could reach its percussion cap only through its allotted hole in a metal disk. A crank turned the disk so that one pin after another lined up with its proper hole and the barrels were fired in succession.

With a good crew the mitrailleuse could deliver the contents of ten magazines in a minute.

124

MONTIGNY MITRAILLEUSE

EARLY GATLING GUN

That was real quick-firing and the gun itself was a success, but circumstances were against it. First, there was so much secrecy about its production that it was kept under wraps too long and gun crews were not properly trained to use it. Second, since it looked like a field gun and since the crews were artillery men, they used it like a field gun. Of course it flopped; it was designed for much closer range. When almost by accident it *was* used at close range, it was fearsomely effective; but that was too late. Another thing that embarrassed the mitrailleuse was that all the secret-weapon build-up caused the Prussians to greet it with the heaviest possible artillery fire as soon as it appeared.

This French gun didn't use metallic cartridges and neither did Dr. Richard Gatling use them in his, which was the first practical American quick-firing gun. The doctor was a go-getter! The Civil War was on and when the Army showed no signs of trying out his gun in the field, Gatling hired civilian crews and gave convincing demonstrations under real battle conditions. After the War the Gatling gun was adapted to metallic cartridges and manufactured by Colt. The Army used it in the Spanish War. An automatic pistol is still a "gat" in American slang.

The Gatling gun was made with ten barrels, each with its own lock, and it was operated by a hand crank but not at all in the way the mitrail-

leuse worked. The Gatling crank rotated the barrels themselves around a central shaft and provided the energy for them to handle the ammunition which was fed to them by gravity from a hopper above the gun. Each barrel received a cartridge when it reached the top position; then, as it moved around from station to station the shell was pushed into the chamber, the breech was closed, the gun was cocked, the shell was fired, then the breech was opened again and the case was extracted and ejected. By then that barrel was ready for another cartridge.

This is a simplified version of the performance which went on simultaneously and continuously, attaining the excellent rate of eight hundred shots a minute with no trouble from overheating because each barrel fired only one shot in ten. It didn't often, if ever, deliver an actual total of eight hundred shots in one minute but it could be fired at that *rate*. This is true of later machine guns too: they are fired in short bursts rather than continuously. In tests, about a thousand successive shots can be fired without pause before the barrel is ruined.

There were other crank-operated quick-firing guns: the Lowell, pretty much like the Gatling; the Hotchkiss, which was a small five-barreled cannon; and the Gardner which was very handy because it weighed only two hundred pounds and

125

GARDNER PORTABLE QUICK-FIRING GUN

MAXIM AUTOMATIC MACHINE GUN

could be packed on one horse. All of these were invented by Americans but were used by European armies.

Hiram Maxim put an end to crank-turning in 1885 when he produced a gun which used its own recoil energy to load and fire itself and eject its own empty shells. This was the first fully automatic gun, the true machine gun. The operator had only to hold the trigger and the gun could continue firing until its ammunition was gone.

Maxim abandoned the gravity-loading idea in favor of a mechanically fed canvas belt in which two hundred and fifty rounds could be packed in advance and the whole thing conveniently boxed. The boxes could be changed rapidly. The Maxim gun had only one barrel which was kept cool by being surrounded with a water jacket. Perhaps the most remarkable feature of this gun, one that is a *must* in high-powered modern machine guns, was that the breechblock was locked tightly to the barrel at the instant of firing. Immediately after firing the recoil moved both barrel and breechblock back together, and as they moved the block was unlocked and continued back after the barrel stopped. As it separated from the barrel, the breechblock took the fired cartridge case along and discarded it.

The Maxim used one of the two principal ways of operating a machine gun—recoil; the Colt-Browning which appeared in 1895 used the other way—gas pressure. A little hole drilled into the under side of the barrel near the muzzle let a minute quantity of gas escape, just as the bullet was leaving the bore. Small as the escaping volume

COLT-BROWNING MACHINE GUN

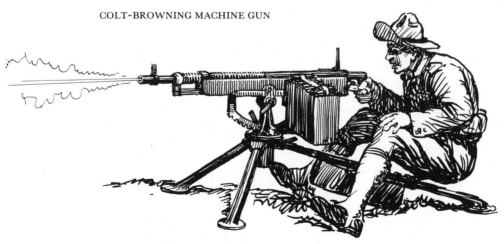

was, it had enormous kick behind it and it hit a little piston head hard enough to furnish power for the whole complex operation of ejecting, reloading and firing again. This gun was the invention of John M. Browning who was one of America's great men.

The first Brownings were air-cooled and were rated at four hundred rounds a minute. They were bored to use the regular infantry issue cartridges, such as were used in the Krag-Jorgenson rifle of the time. Later, a heavier water-cooled Browning was built but the air-cooled type was used through World War I, and was one of the guns to which special controls were applied to make them shoot between the blades of an airplane propeller without hitting it.

## Shoulder Arms and Hand Arms (1900–1925)

Shotguns using paper-jacketed shells became very popular around the turn of the century and some very fine ones were built. Many were double-barreled, most with the barrels set side by

U. S. AIR FORCE SURVIVAL GUN

side, a few with one barrel set over the other. Guns have been made (and still are) which have one shotgun barrel and one rifle barrel; these are usually the over-and-under type. The little survival gun now carried by American fliers to kill small game for food is an over-and-under rifle-shotgun. Ammunition for both barrels is carried in the stock.

The sizes or *gauges* of shotgun bores are based on the ancient system of sizing a musket barrel by the number of bullets a pound of lead would make for it. The largest shotgun is a four-gauge "elephant gun" which throws a quarter-pound chunk of lead; the smallest shotgun is a lady's twenty-gauge, with a bore not much bigger than a lead pencil.

In spite of being measured by the size of the solid ball that will fit them, shotguns are used for firing charges of small shot which spread out from the muzzle and form a *pattern* or *killing circle* for small game. This pattern should be about thirty inches in diameter when the shot is fired from a twelve-gauge gun at a distance of ninety feet. If the bore is the same size all the way out, the shot will be evenly distributed in the circle; if the barrel is "choked," that is, constricted a little near the muzzle, there will be a concentration of shot near the center of the circle.

Up to the middle of the eighteenth century shot was made by molding for the larger sizes and cutting up sheet lead for the small sizes. Then it was found that molten lead poured through a sieve and allowed to fall from a high tower would form itself into neat, round balls and harden as it fell. The shot were caught in a tub of water and sorted

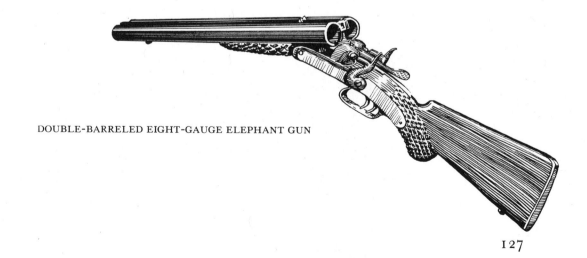

DOUBLE-BARRELED EIGHT-GAUGE ELEPHANT GUN

127

for size later. One or two of the old shot towers are still standing in Atlantic seaboard towns.

Shot comes in many sizes, from large buckshot, five of which will make an ounce, down to Number Ten, of which it takes 1600 to make an ounce. Once each size had a name; there were swan drops, goose drops, duck drops and dust shot. The last was half the size of Number Ten. Small sizes are still made by dropping, but buckshot and larger sizes are cast in molds.

Repeating shotguns had proved practical almost as soon as the first revolver had, and it wasn't long before a fully automatic shotgun appeared. Sporting rifles, too, progressed early in the twentieth century from repeaters to automatics. It was not difficult to make an automatic as long as the charge wasn't too powerful. This fact, plus the fact that no infantry gun may weigh more than nine pounds, is what so long delayed the military automatic rifle.

The first automatic pistol made in America was the Colt .38 introduced in 1900. It had an eight-shot magazine which slipped into the butt just as later Colt automatics do. The work was done by the recoil of the barrel. The caliber was increased to .45 in 1905, and in 1911 the gun was adopted as the official U.S. Army side arm. The Colt is the most dependable of all automatics, an important characteristic since it is intended for personal protection at close range and close range is no time for a gun to jam or miss fire. It's quite a gun. It will stop a running man in his tracks and will flip a light man clear over. It delivers a mighty wallop and it jumps in the shooting hand like a fresh-caught salmon. This makes it a dangerous weapon for bystanders.

THE FIRST COLT AUTOMATIC PISTOL

The faults of the Krag rifle were such that no modifications could ever remedy them. So the Army quit fooling with it and designed the .30-caliber Springfield M1903 which was based on the German Mauser. Slight changes were made later, but by and large the same great rifle served our armed forces through the first World War and well into the second. The Marines took their Springfields onto Guadalcanal and they still use them as *sniper* rifles!

The Springfield has won matches against the fanciest target rifles in the world, and yet on other occasions, with her works half-choked with mud and grit, she has seen what was needed and gone on shooting. Like the Krag, this is a bolt-action gun. The bolt is a steel cylinder containing most of the mechanism needed to make the gun shoot. A knobbed handle projects from the right side of the bolt, by which it is moved forward and back in its guide and with which it is locked upon a cartridge in the chamber. With the bolt all the way back, the contents of a five-shell clip can be put into the magazine from the top. Pushing the bolt forward will carry the top cartridge into the chamber. Rotating the bolt by bringing its handle down to the right locks it.

After firing, the bolt is opened by being rotated to the left and drawn back by hand. This action extracts and ejects the fired shell and at the same time cocks the gun for the next shot, the cartridge for which comes into position just when the bolt is clear back.

With standard ammunition the M1903 has an extreme range of 3300 yards; 2200 yards more can be made by using a more powerful shell and a boat-tailed bullet. This is considered to be beyond the practical needs of an infantry weapon. The same idea stands against the experts, who have argued that this gun is better than its sights and that with better sighting devices its effective combat range could be increased beyond the thousand yards which is now given for it. There's room to doubt that in any of our wars John Citizen Doughfoot will ever have time to train for the use of such a sight, or much opportunity to use it if he knew how.

The M1903 is a full-stocked gun like the old Kentucky rifle; the wood is carried nearly to the muzzle and clear over the top of the barrel. But

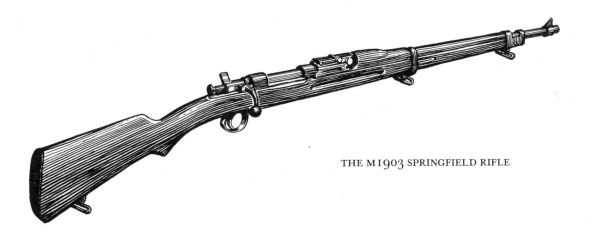

THE MI903 SPRINGFIELD RIFLE

the gun is a good two feet shorter than a Kentucky, just over forty-three inches long, and it weighs less than nine pounds. Originally this Springfield carried a sixteen-inch knife bayonet, but in late years this has been shortened to ten inches.

The first suggestion of a submachine gun, which would fall between a machine gun and an automatic pistol, came with the "Pederson devise" secretly developed in World War I. Replacing the bolt of a Springfield rifle, this gadget would make the gun capable of delivering an astonishing spray of .32-caliber pistol bullets effective up to a hundred yards and very useful for close fighting. The thing was due to be tried in France in 1919 but the Armistice was signed before it was combat-tested or even revealed. All the units which had been made were destroyed after the

war to keep them out of the hands of the lawless. This seems to have been a futile gesture, since the Tommy gun, which was the real answer to the gangster's prayer, appeared shortly and evidently was available in quantity.

From 1900 to the beginning of World War I there was a rash of new machine guns in all nations. Some were excellent weapons but they added little that was new in the way of principle. One worth at least passing attention is the Lewis, invented by an American but adopted by the

THE LEWIS LIGHT MACHINE GUN. The aluminum cooling fins are covered by the steel jacket

129

British as their light machine gun in 1916. It was an air-cooled gun, the barrel being covered with a finned aluminum sheath to carry off heat. The magazine was a rotating drum mounted on top of the gun. The Lewis was simple in design and easy to tear down in the field when it jammed, which was frequently .

For air use this gun was swivel-mounted with the barrel bare, and a "ring" sight was mounted on it. This trick helped an aerial gunner to make some estimate of how far ahead of his moving target he should point the gun in order to have some hope that bullet and target would reach a given point at the same time.

The development of tracer bullets which showed their path as a streak of light was a great help to the success of plane-to-plane gunnery. It made shooting more like spraying a hose. Incendiary bullets were mixed into the magazine too. Since the planes of 1918 were nothing if not inflammable, these bullets often did the trick when plain bullets failed.

## Great Guns and Little Guns (1900—1925)

This is about the artillery of World War I, or at least some of it; this war was the proving ground for all that had been thought up since 1904. Of all the changes, the greatest one was from horses to tractors for moving guns. There are some bow-legged gentlemen who haven't yet recovered from that blow!

Artillery was broadly divided into two classes, fixed and mobile. The more powerful it was the more fixed it became. Mobile artillery included anything from the little 37 mm. infantry cannon which two men could carry, up to the enormous railway guns, some of which were nearly sixty feet long and which actually belonged to the class of fixed guns except for the bare fact that they could be moved—a little. Fixed guns no longer have any military use.

Starting with the little fellows: World War I 37 mm. gun was an infantry weapon mounted on a tripod like a machine gun and served by a two-man crew. These two, when there was need to move, simply divided the gun and the mount between them and carried. The whole object of the gun's design was to make it as light and movable as possible for its caliber. It was used against "points of resistance" which were chiefly machine-gun nests. An explosive shell an inch and a half thick and weighing a pound and a quarter is not a comfortable thing to have with you in your little machine-gun nest.

TRUCK-MOUNTED ANTIAIRCRAFT GUN 1918

37 MM. GUN ON A TRIPOD MOUNT. The cone at the muzzle is a flash-hider

When "aeroplanes" began to be a factor in war, it was necessary to find ways to shoot at them with artillery in the field. One of the ways was to mount the old "75" with its breech hung over the tailgate of a truck, so that it could fire at a very high angle. Shooting at moving aircraft was a new experience for the artillery. Planes weren't very fast in those days, but they were too fast to allow time for intricate calculations of range and so on. You aimed a "75" a little ahead of a plane, much as you aim at a flying duck with a shotgun, you "led" him and fired. The surprising thing is that some airplanes were brought down in this way.

Recently the Army announced the "Skysweeper," an antiaircraft "75" which has its own radar, able to control the aiming and the *firing* of the gun. It will scan the sky tirelessly for enemy aircraft, find one flying better than five hundred miles an hour and track it from fifteen miles away.

When the plane comes within four miles, the Skysweeper aims its gun continuously, not at the plane but at the swiftly changing spot where the plane will meet a shell fired at any given instant. This the gun does in darkness or fog as readily as in clear daylight, computing speed, course, drift and range, and firing and reloading automatically like a machine gun, though slower because of the size of its shells. These shells have proximity fuses which explode them as they approach the target, instead of when they actually hit it.

The "75" on its regular carriage was surely the most-fired cannon of World War I. Along with the 75 mm. howitzer and the 105 mm. howitzer, it *was* the light artillery, operating just behind the front line trenches. Howitzers fired at a higher angle than guns but not so high as mortars. They still do, but the distinction is fading now because all guns except trench mortars have to fire at *any*

131

angle. The Skysweeper can be turned against tanks on the ground if the need arises.

Medium artillery consisted of the three-inch gun and the 155 mm. howitzer. This howitzer we adapted from the French Schneider. It was pulled on the move by a five-ton tractor. Like any field gun it had a trail which rested on a two-wheeled limber. Both the gun carriage and the limber had solid rubber tires. Along with the howitzer went its own light repair truck, its ammunition cart or carts and a big reel on wheels, for telephone wire. The howitzer cannoneers were never able to see their target; they had to take firing information by phone from a central control point.

The 155 mm. howitzer could lean back and heave a shell 12,530 yards and it could do that twice a minute, even faster when the pressure was on. As with all guns except the smallest, three types of shells were used; high explosive, shrapnel and gas. These were *separate loaded* in the 155 and larger sizes, that is, the shell was put into the gun and the packaged powder and a primer were shoved in behind it.

For the really heavy work there was the 155 mm. gun and two big howitzers, an eight-inch and a 240 mm. (nearly nine and a half inches). The 155 mm. gun was probably the most mobile of the heavies but it was still unquestionably heavy. With its carriage it weighed 25,960 pounds. A big tractor could roll the gun along a decent road at *eight* miles an hour. World War II guns usually moved at fifty or sixty.

All of these heavies were used to "lay down" the barrage which the French invented to cover an infantry advance. A heavy concentration of artillery fire was thrown on the enemy trenches. At the "zero hour" infantry moved against these trenches and the range of the big guns was increased just enough to keep the shells falling ahead of the advance.

World War I was a railway gun war: it stayed put long enough for the things to be set up. They could shoot twenty miles but they had to be run on tracks even though they didn't actually fire from them. The special flat cars on which the guns were mounted were jacked up and solid bases were laid under the guns before they were fired. In World War II the Germans fired two railway guns known together as "Anzio Annie," but no American railway gun fired a shot.

Though it was important only as propaganda, the German's Paris Gun is interesting because it still holds the record for plain long-distance shooting—seventy-five miles, no less! Actually it was not one gun but several, because the pressures needed for what they were doing ruined them after fifty or sixty rounds. A lot of people call this gun "Big Bertha," but they have the wrong girl; Big Bertha was a sixteen-inch howitzer the Germans used in Belgium in 1914.

The Paris Gun was made from a fifteen-inch naval rifle by inserting a tube which reduced the bore to eight and a quarter inches and nearly doubled the original fifty-six-foot length. The fin-

FOURTEEN-INCH RAILWAY GUN IN FIRING POSITION

THE PARIS GUN

ished barrel was so long that it drooped almost an inch and had to be held straight by a truss. The 250-pound shell took three minutes to get to Paris from the Forest of Saint-Gobain, east of Soissons. The gunners had to do some dizzy mathematics to allow for the forty miles the earth rotated while the shell was in flight.

In 140 days of intermittent firing (a shell every twenty minutes when things were going good), 256 Parisians were killed while several hundred thousand others lived in a state of frantic jitters. Those who are old enough to have seen it will never forget the shock of the headline, GERMANS SHELLING PARIS.

The long sixteen-inch rifles and the big howitzers and mortars which defended our coasts for the first half of the century are for the most part still in place, but they are not so impressive and comforting as they once were. Naval armament and accurate bombing from carrier planes have made the "ramparts" look pretty silly, and guided missiles aren't going to add a thing to their importance.

Let's take a little closer look at twentieth-century "cannon balls." The standard is the high explosive shell which may be thin-walled to hold a maximum bursting charge for blowing things up, or thick-walled so that when it bursts, many le-

133

thal fragments will be scattered. There is some kind of fuse; it may be a time fuse to burst the shell after a predetermined number of seconds; it may be a point-detonating fuse which explodes the shell instantly when it touches anything, or which gives the shell a second or so to penetrate before it goes off. The bursting charge is usually TNT which looks a lot like brown sugar but acts quite differently. If the shell is intended to pierce armor it will be almost solid steel with a hard, blunt head hidden under thin streamlining.

Just back of the head of a shell (any shell) there's a narrow band formed on the steel itself, which projects just a very little bit from the body of the shell. This is the *bourrelet;* its function is to guide the front end of the shell in the bore and it's machined very exactly to fit the bore with just enough play, five thousandths of an inch in small

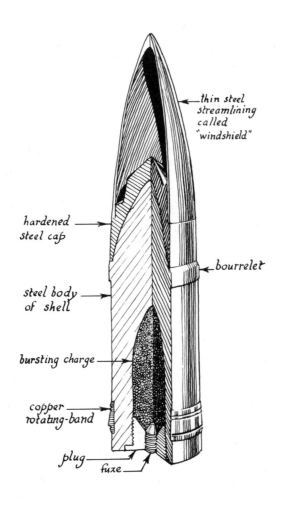

ARMOR-PIERCING SHELL WITH A PIECE CUT OUT TO SHOW THE INSIDE

134

guns. Around the bottom of the shell and shrunk to a secure seat by being put on hot is the soft rotating band, just a little *greater* in diameter than the bore of the gun. Its sides are not parallel with the body of the shell but are tapered slightly towards the nose. The angle of the taper exactly matches the one in the bore at the beginning of the rifling, against which the rotating band lies when the shell is ready to be fired.

When the shell passes through the bore the only parts of it which touch the gun are the bourrelet and the rotating band. Thus the band "locates" the shell, supports its base and seals off the forward leakage of gas by filling the rifling grooves. Sometimes more than one band is used.

The shells fired by very large guns are much too heavy for men to lift. They are handled with tackle as shown in the sketch of the railway gun. For loading, the gun barrel is depressed about five degrees to let the projectile slide in. Impelled only by that slight incline, the rotating band hits the beginning of the rifling so hard that the grooves cut into the band and will hold the shell where it is, even if the gun is elevated.

A shrapnel shell is a blunt-nosed steel can, actually a gun in itself, though it is fired from a gun. There is a charge of black powder in the base of the shell; the rest of the space is crammed with lead balls embedded in rosin. The shell is shot towards its target like any other projectile. Just as it arrives, a pre-set time fuse fires the black powder. The shell doesn't break up. Instead it acts as a mortar, and the lead balls are discharged from it exactly as shot are fired from a shotgun. Shrapnel was much used in World War I but none was used in World War II. What were often called "shrapnel wounds" were actually caused by fragments of shells or grenades.

The latest "scatter shot" just coming into use is a revival of the old canister invented by Gustavus Adolphus and important in the American Civil War. The modern canister is intended for use at very close range against infantry. It is fired from rifled guns and so comes out spinning. Leaving the gun muzzle the case opens out, and the spin spreads twelve hundred or so pellets like buckshot. It is very effective when a mass of infantry is too close to be handled by machine-gun fire; it is also used to clear invisible or merely suspected

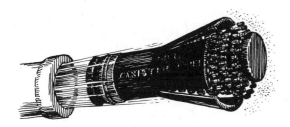

CANISTER, 1953

snipers out of trees, or even for tearing away foliage so that gunners can see where to shoot.

Chemical or gas shells, and incendiary shells, are built like high explosive shells except that they have only enough bursting charge to open them up, the rest of the space is given over to the chemical material. There are also "star shells" whose only purpose is to burst high above a target at night and cast a brilliant light on it, so that other shells may be sent to the right address.

## Special Weapons (1900—1925)

Leonardo da Vinci planned a "secure and covered chariot with guns" but it is not known to have been built. An English plumber suggested a "land cruiser" in 1911. He was promptly brushed off, but his idea was dug up a few years later when something had to be found to break the deadlock between the armies dug into the French mud. Winston Churchill got behind the plumber's idea with enough push to make a reality of it, even against the disgusted resistance of the top British brass.

To keep the secret of what was really being tried, the "armored machine-gun destroyers" were referred to as "tanks," so tanks they remain. The first slow, clumsy behemoth they named "Mother" and the second was "Little Willie." Though at first nobody understood how to use them properly, these monsters which crawled across trenches and through shell craters, knocking over trees and spitting bullets as they came, nearly wrecked the morale of the German army. In time Fritz learned that a grenade in the caterpiller track would put a tank out of action, and that a pit too wide to bridge and too steep to climb would trap the brute; but he never learned to feel really comfortable about tanks.

Late in the war the Allies sent their tanks over in big bunches behind heavy barrages. A hundred or more of the things together were at that time irresistible. It was such massed tank attacks that began to turn the tide in 1918. The Germans built tanks too, but only halfheartedly; they had just fifty at the end of the war. By then the British had heavies that would cross a ten-foot ditch, and also light, fast "whippets."

Military men have been accused of being always fully prepared to win the war that has ended but never ready for the next one, which will be fought quite differently. Yet if tanks win a war, we aren't safe without more and better tanks for the next war and we should certainly develop every possible defense against tanks; even though we make

DA VINCI'S "TANK"

BRITISH HEAVY TANK IN WORLD WAR I

135

PINEAPPLE HAND GRENADE

an effort, at the same time, to go a step further with some other weapon which will itself dictate what kind of war the next is to be.

In the war against the Kaiser the new power of the machine gun pinned both sides to the mud for months on end. Since nobody who could help it ever put his head above ground, weapons had to be found which could be lobbed high and dropped into trenches almost vertically. The "pineapple" hand grenade was just the article for the infantry in such a case. It was made with a cast-iron body deeply cut by grooves crossing each other at right angles, and it hasn't noticeably changed. The grooves are to make sure the body breaks up well because in a fragmentation grenade the pieces of the body are the "shot."

A grenade is just a nice size to fit in the hand —it hefts well, like the savage's throwing stone. Along one side of the pineapple lies a long trigger which a spring tries to force away from the body of the grenade but which is held safe by a pin. Once the pin is pulled, the grenade is *armed*

and the trigger must be *held* down, or the grenade will go off in the thrower's hand.

Of course the trigger springs out as soon as the grenade is thrown, and its business end sets off a short fuse which explodes the pineapple five seconds later. It's just as well to heave a grenade from behind a bank or a tree because it will scatter its forty metal slugs in all directions and just about as far as a man can throw it; you *could* find yourself on the receiving end of your own weapon. Grenades thrown at close range have been grabbed and thrown back to the sender before they could explode. Anyone who plays this kind of ping-pong should work fast and think about something else afterwards.

Not all World War I grenades were of the pineapple variety. Some were long-handled "potato mashers" which were supposed to have added range. Some were fired from rifles. There were two kinds of rifle grenades. One was started on its way by firing a regular rifle bullet right through it. The other kind carried a steel rod ten inches long which was inserted in the rifle bore. A special blank cartridge pushed it out of the barrel and threw it about two hundred yards. Neither kind did a rifle any good and both were abandoned after the war, but the idea has been revived and an improved grenade launcher is now issued as a rifle attachment.

Trench mortars accomplished the same purpose achieved by grenades, but they carried more stuff further. A trench mortar is usually considered a gun, but the original ones seem more closely related to a gas pipe. They weren't aimed. Like the earliest medieval cannon they were merely pointed in the general direction of the target and cut loose. To fire one you simply dropped the projectile into the muzzle and snatched your hand

RIFLE GRENADE AND LAUNCHER

136

CURRENT MODEL OF 60 MM. TRENCH MORTAR

mortar, though it is muzzle-loading, has a rifled barrel and a range above 4000 yards.

Some special weapons were developed for use at sea. The torpedo is associated in our minds with submarines, in fact the pigboats were originally called "submarine torpedo boats." Actually the idea of sneaking an explosive charge up to the hull of a ship and setting it off is a lot older than the first successful submarine. Something of the sort was tried in the American Revolution and Stephen Decatur accomplished it when he blew up the *Philadelphia* in Tripoli harbor. In the Civil War the Navy had torpedoes attached to long spars, which were floated up to an enemy ship at night by a skiff or even by a swimmer.

The torpedoes which were launched by and at the German U-boats were all of the Whitehead type which were self-propelled and automatically steered. The nose or "war head" held five hundred pounds or so of TNT behind a delayed-action fuse which allowed the "fish" time enough to penetrate a hull before it exploded. A torpedo was likely to hit hard enough to make more than a dent, too. Even in early days their little steam engines drove them along at twenty-five knots. The steam was generated by burning alcohol with compressed air.

A torpedo is launched by being pushed out of a big tube by compressed air. The steering mechanism in the tin fish is gyroscopically controlled and is pre-set to guide the torpedo to its target. The most recent torpedoes have been provided with a device which makes them *seek* the target. They will actually change their course to follow a ship which is trying to evade them.

The success of the German U-boats in both wars has demanded the invention of countermeasures. The first problem was finding the pesky things. Radar is helpful but listening devices have proved better for this—they have been brought a long way from the first crude instruments to the present highly sensitive *sonar*. Once located, some fathoms below the keel of an attacking surface ship, the next poser was how to attack. Guns were no good, nor was anything which had to be accurately aimed at an invisible target. The answer was found in the "ash can," more properly but not often called a *depth bomb*. It will damage a sub even if it goes off twenty-five or thirty yards away,

out of the way. There was a fixed firing pin at the bottom of the pipe, and when it met the primer of the twelve-gauge shotgun shell which served as the "propellant," the projectile came back out again and quick!

The shotgun cartridge by itself would throw a shell three hundred yards. By boosting with added *ballistite* a maximum range of 750 yards could be reached. Ordinarily a three-inch trench mortar could fire ten times a minute but the speed could be stepped up to twenty-five a minute in a pinch. Most trench mortar shells were high explosive projectiles weighing from seven to ten pounds. Modern trench mortar shells, except the largest, have four steel fins to keep them on course, but the 1917 variety were simple "cans" equipped with a detonating fuse which would go off even if the shell landed wrong end up.

Three sizes of trench mortar are now used by the Army. The two smaller sizes haven't changed too much from the old ones, but the M2 4.2-inch

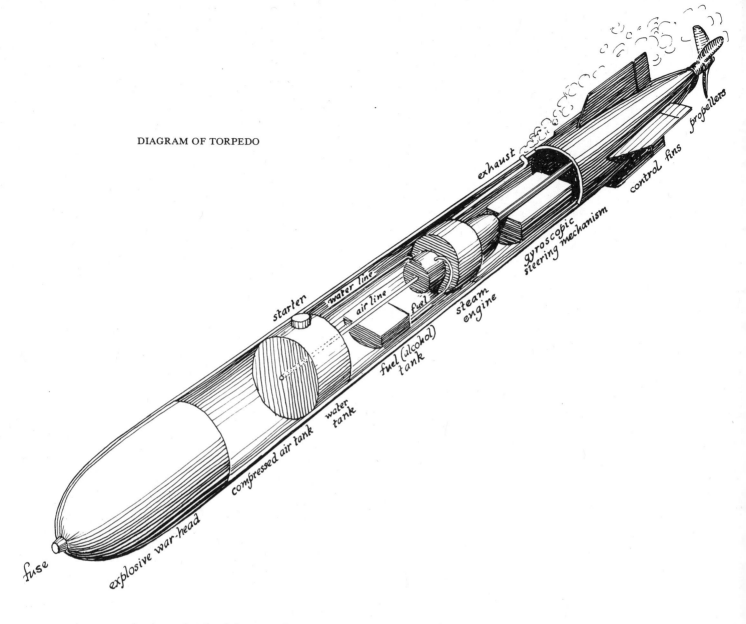

fuse

explosive war-head

compressed air tank

water tank

starter

water line

air line

fuel (alcohol) tank

fuel

steam engine

gyroscopic steering mechanism

exhaust

control fins

propellers

because the shock is transmitted through the water. The system is to try to surround the sub with a "pattern" of depth-bomb explosions, set to occur at the estimated depth of the target. Dropping ash cans from aircraft proved very effective against the submarine wolf packs of World War II.

Stationary submarine mines were originally known as torpedoes. Up to the end of World War I, they fell into two simple classes; controlled and self-acting. Both kinds floated at the end of an anchored cable to any desired depth, usually about fifteen feet. Mines are still laid in groups or "fields" in harbor entrances. The controlled kind are connected to the shore by electrical wires and explode only when a switch is thrown.

The self-acting or contact mines have largely replaced the controlled kind but they have the disadvantage of not being choosy. Some of them are fitted with a number of projecting triggers any one of which will explode the mine if it is touched by a ship; and it doesn't care what ship. On the more recent mines the triggers are thin glass bulbs which have only to be broken to set off a hundred and fifty pounds of TNT. Other modern refinements are the magnetic mine which is exploded by the magnetic field of a ship passing near it; and the acoustic mine which "listens" and goes off when it "hears" a ship's propeller nearby. Neither of these requires actual contact.

The Navy has lately admitted the existence of what may be the strangest "weapon" of them all: the "frogmen." These fellows are good at swimming. They wear special waterproof suits and fantastic rubber flippers. With an hour and a half's oxygen supply tanked on their backs, they can leave a submerged submarine and return to it without ever going to the surface. One of their main jobs is clearing mine fields by personally removing the mines.

In addition to the sea mine there is the land mine. It's usually buried by retreating troops and depends on pressure from above to detonate it. The weight required may be only that of a man's foot or it may be set to be affected by nothing lighter than a truck or a tank. An instrument which looks like a pancake-on-a-stick will detect the presence of a metal mine underground and signal its operator. Before this was invented, the method of finding shallow mines was to drive a herd of pigs ahead of advancing troops.

Land mines make good tank traps. Even if one does no more than blow off a tread, it stops the tank. The Maginot line along the border of France was equipped with short concrete posts for tipping tanks over, but these never got a chance to tip any tanks because the Germans decided to come in the back door which no one had bothered to bolt; they had come that way before, but no one thought they'd do it again! The banked hedgerows which surround fields in Normandy turned out to be completely effective tank traps until an American soldier made a plow attachment for our tanks, with which they could cut right through the banks.

NAVY FROGMAN CLEARING A MINE FIELD

MINE DETECTOR

WORLD WAR I GAS MASK

ably uncomfortable, protection could be had from a mask which filtered air through charcoal and chemicals. Later, tear gas, mustard gas and other skin irritants were introduced. All gas is abominable and very tough to control. A change of wind and you're attacking yourself with your own weapon! It was this, plus the fact that both sides were ready with it, that kept gas out of World War II.

Of all the special weapons of World War I, the most spectacular was poison gas; but its actual accomplishments as a weapon left something to be desired. It was sprung by the Germans. At first they merely released it from containers and allowed it to drift on the breeze; later it was loaded into shells and grenades. The first gases used affected the lungs only and a reasonable, if miser-

## Self-Loading and Automatic Guns after 1925

The Springfield was a repeater; a fired round could be removed from it and a new one thrown into the chamber by opening the bolt by hand and closing it again. The Garand is a self-loader or semiautomatic; it removes its own fired shell and puts a new one into the chamber, getting power to do it from a little gas spurting from a hole in the barrel near the muzzle and moving a piston in a cylinder under the barrel. When its

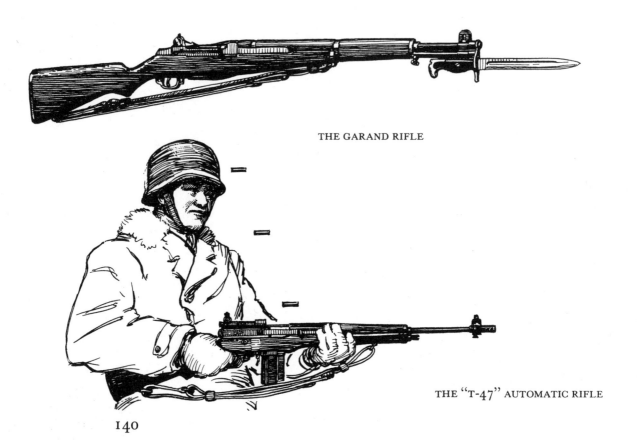

THE GARAND RIFLE

THE "T-47" AUTOMATIC RIFLE

THE M I CARBINE

cycle is complete, the gun is cocked and ready to fire again but it doesn't fire itself. The trigger must be pulled by hand and released after every shot.

The Garand magazine holds eight rounds which a trained soldier can fire with reasonable accuracy in twenty seconds. The same man will take twenty-five seconds to fire the *five* rounds of a Springfield with the same accuracy. The Garand was officially adopted as the United States infantry weapon in 1940 and amply proved itself in Europe and in the Pacific. General Patton called it "the greatest battle implement ever devised."

A fully automatic rifle is now being tested. Like a machine gun, it will continue to follow one bullet with another as long as the trigger is held. It is about a pound lighter than the Garand and as accurate in semiautomatic fire (it can work both ways). The objection to a fully automatic infantry rifle has always been that it would waste ammunition and that an excited soldier might fire himself defenseless; then too, it's hard to hold a gun on its target when it's popping bullets at a high rate: the muzzle will climb.

The new rifle fires a bullet of the same weight and striking force but its cartridge is shorter than the one used in the Garand. It is this shorter cartridge which has made it possible to build the "T-47" lighter than the older gun. The magazine of the test model holds twenty rounds. This gun shows up well in trials but it hasn't yet been adopted by the Army. If and when it is adopted, it will be some years before it replaces the Garand.

The beloved Colt .45 automatic pistol is so much a part of all the American forces that it's startling to realize that it has become obsolete and has already been extensively replaced by a very different weapon, the pistol-that-looks-like-a-rifle,

the M1 carbine. Once carbines were short rifles, though some were made with pistol grips and detachable shoulder stocks. One model of the M1 is made that way for paratroopers. Originally a pistol was planned as a one-handed gun to make it easy to shoot on horseback. Very little riding is done in modern armies and any shoulder weapon shoots straighter than a one-handed one, particularly in the hands of American citizen-soldiers who have seldom managed to master a pistol. It's been whispered that the great .45 has never been proved to have killed an enemy in combat.

A special, short caliber .30 cartridge has been built for the carbine's shorter chamber. It has a lot less zing than the regular rifle cartridge but considerably more than the .45 automatic ammunition. The carbine can hit things three hundred yards away but its ideal range is anything up to half of that. The gun weighs only five and a quarter pounds. It is semiautomatic like the Garand or fully automatic, by the firer's choice. The magazine holds fifteen rounds.

This weapon is being used by the many special troops which an army now needs, men whose normal duties have nothing to do with guns but who sometimes need a "shootin' arn" and need it bad. In the past engineers and such have shown a marked tendency to "lose" rifles when carrying them became a nuisance. The carbine is light enough to encourage such men to keep it with them and easier than a .45 for them to handle successfully. It makes a good officers' weapon too, and it seems to be a comfort to paratroopers who can jump with it and land shooting, instead of waiting for a gun to come down on the next elevator.

The automatic carbine with a "sniperscope" added becomes a formidable weapon for night

THE AUTOMATIC M2 CARBINE WITH
"SNIPERSCOPE"

fighting. A soldier using it can literally see in the dark without being seen. He sights his gun through a special 'scope mounted on top of the barrel. Through it he sees by "black light" (infrared) which is invisible to the unaided eye. He has his own battery-operated "searchlight" mounted on the gun under the barrel. Of course if the enemy has the right glasses, the sniper becomes a singularly conspicuous target.

Submachine guns are bullet sprayers which handle .45-caliber pistol bullets. They don't shoot any farther or harder than an automatic pistol but they shoot faster and they can stay at it longer without reloading. The Thompson submachine gun has been found to work better and waste less with a twenty-shot magazine but it can be used with a fifty-shot drum. The Tommy gun has an excellent and honorable record for performance in jungle fighting; it also at one time fell into bad company and was used by competing "businessmen" to drill neat rows of holes in one another.

Submachine guns operate by a system known as *blowback*. The principle of this is based on the

fact that it's easier to move a light object than a heavy one. When a cartridge is fired, its gases push forward against the bullet and back against the breechblock. In the blowback system the breechblock isn't locked but it's heavier than the bullet, so the bullet moves first. By the time the breechblock starts, pressure in the bore has dropped enough for the breech to be safely opened. Straight blowback will work only on light arms; heavier weapons like the Tommy gun, use what is called retarded blowback: some kind of drag is added to the breechblock to give it extra resistance. Blowback absorbs quite a large part of the kick of a gun.

The remarkable Finnish Suomi submachine gun which has only one moving part, the bolt, is hinted to have been built upon an American gangster invention which crept into Europe by way of South America. The Russians have taken the Suomi to their hearts because it works well in cold weather.

A simple, light "machine pistol" which is tough but cheap to make has uses in a modern army. Many of the uses are the same as for the M1 carbine, except that a submachine gun gives greater firepower within its more limited range. The British came up with the various forms of the Sten gun which, it was said, could be "made in any garage." The Americans took long looks at the Sten and the Suomi and then issued the M3.

The M3 is a cheap submachine gun. It isn't handsome. Its shoulder stock is nothing more than

TOMMY GUN

a piece of heavy wire with a crook in it. Most of the gun's parts are stamped out of sheet metal the way pots and pans are made, yet it's fully automatic and will spit slugs at the rate of 450 a minute. Its long, narrow magazine holds thirty rounds. The effective range of the M3 is considered to be a hundred yards but it can actually throw a bullet farther than that.

The need for light, handy machine guns which could travel with the infantry was discovered in World War I and resulted in the Lewis, the light Browning and the French Chauchat. Such guns are still needed and they have been made even lighter and more handy, so that they approach the automatic shoulder rifle and may eventually be eliminated by it.

The heavy machine gun has been made much heavier. The old .30-caliber has been replaced entirely in the air and largely on the ground by the .50-caliber guns. The Germans invented those big fellows in an effort to stop British tanks in World War I. In World War II the tanks had gone beyond the gun but it still did well against trucks and as an antiaircraft gun.

The standard .50 caliber in the American forces is the Browning. Mechanically it is very like its older .30-caliber brother which spat lead thirty years before Pearl Harbor. Some of the fifties are

THE M3 SUBMACHINE GUN, known as the "grease gun"

water-cooled but mostly they are cooled by air and the barrels are changed when one gets too hot. Ammunition is fed from hundred-round metal link belts. These are held together by the cartridges themselves and fall apart as the rounds are fed into the gun. There are differences between Brownings for ground use and those for the air, but they are only the minor differences necessary to adapt each to the conditions of its service.

Above all the Browning .50 has done well as an aircraft-mounted weapon. By the war's end no American or British plane carried anything lighter than a fifty. Some were set inside wings, some were set on movable mounts, some were synchronized to shoot between the blades of propellers and some actually fired through the hollow crankshafts of engines.

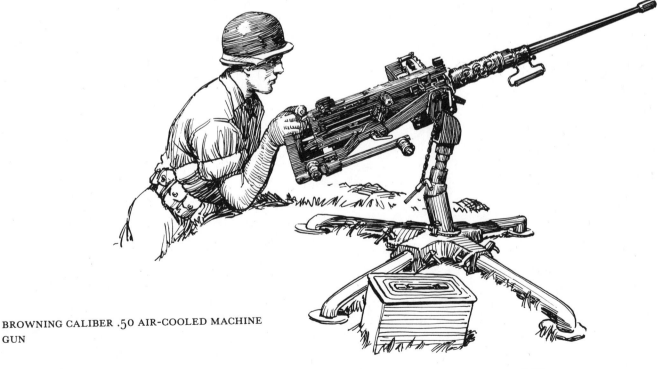

BROWNING CALIBER .50 AIR-COOLED MACHINE GUN

143

Some quite big guns have been tried on planes, cannon of .37 mm. and even 75's. These have great stopping power if a hit can be scored with them, but their action is too slow for much success in the air except for dive-shooting. Several Japanese ships were actually *sunk* by 75's used in this way. Because of its faster action a 20 mm. cannon will throw more metal per minute than the larger guns. This is the size which has proved most effective for air-to-air use because it is small enough for really rapid fire and yet large enough to use an explosive bullet.

When you shoot at a very fast moving target, say one going above three hundred miles an hour, you have a better chance of scoring hits if your bullets come close together. A 20 mm. will give you nine or ten shots a second, a 50 mm. will give you fifteen a second; if you have eight 50's shooting together, you can throw a hundred and twenty-eight rounds a second, which takes a little dodging.

COMPARATIVE SIZES OF .30 AND .50 CALIBER ROUNDS

## 1954

The second World War moved with speed, swinging across great stretches of land and sea. Guns had to go anywhere a road went and a lot

of places it didn't. The tendency of the time between the wars was to make all artillery, even quite big stuff, more rapidly movable. The 155 mm. gun is an example. On its original solid-tired carriage it could be towed at eight miles an hour by a caterpillar tractor. Just before World War II a new, balloon-tired carriage was designed which could be towed at twenty-five miles an hour. During the war the same gun was given a self-propelled mount which could go as fast, turn on a dime and was ready to shoot on very short notice.

These self-propelled mounts are a rational outgrowth of the effort to increase the mobility of

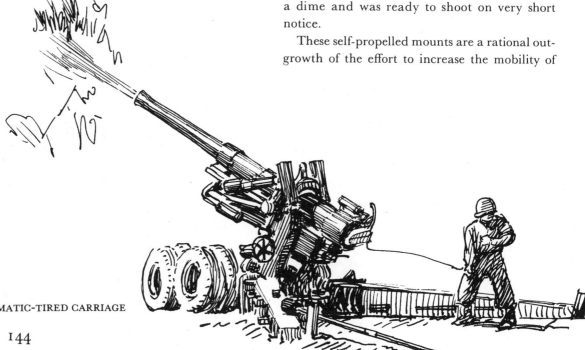

155 MM. GUN ON PNEUMATIC-TIRED CARRIAGE

guns, added to experience in tank design. Most of them have caterpillar treads and John Q. Citizen may well glance at one and say, "Tank," but they aren't tanks—they're self-moving gun emplacements, not intended to give such close support to infantry as tanks do.

The "General Sherman" thirty-five-ton-tank which the United States built in World War II was able to defeat heavier and more powerfully armed German tanks, not only because it outnumbered them but because it was easier to handle and was mechanically sturdier. General Patton, who knew a thing or two about the breed, said that if he had tried his dash across France with German tanks, *all* of them would have broken down by the time he reached the Moselle River. Speaking of rivers, all U. S. tanks are now built to cross any river four feet deep or shallower, and with special gimmicks can cross rivers up to *nine* feet deep.

Late in the war we brought out the "General Pershing" tank which mounts a 90 mm. gun. After the war the Pershing was improved into the faster and more easily handled "Patton" or M46. This one and the Sherman have proved easily able to

rough-up the Russian thirty-five-tonner which gave our lighter equipment trouble early in the Korean fracas.

During World War II Army Ordnance started a plan for standardizing parts of motor-driven vehicles. The same "components" can be used in several tanks and also in a troop carrier, a cargo

"GENERAL PERSHING" 45-TON TANK INTRODUCED IN WORLD WAR II

145

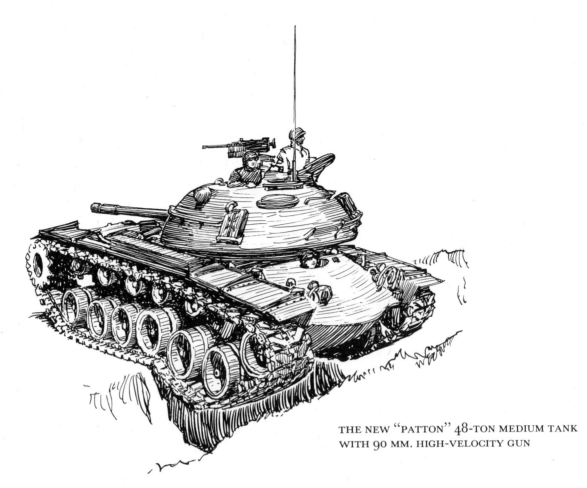

THE NEW "PATTON" 48-TON MEDIUM TANK
WITH 90 MM. HIGH-VELOCITY GUN

THE WALKER BULLDOG 26-TON LIGHT TANK

carrier, a supply truck and a motorized gun carriage. For instance, a single kind of engine serves eleven types of tracked vehicles. This not only saves money for the taxpayer, it also eases the job of keeping these vehicles operating in the field.

For ordinary folks the term "weapons" must now be stretched to cover a much wider range of items than it once did. It must include such things

as the amphibious "Otter" which can carry troops at thirty miles an hour on land and without pause can ferry them across any river that happens to get in the way; it must include the "Eager Beaver," a two-and-a-half-ton truck which can cross the same rivers by traveling on their bottoms, both driver and engine breathing through tubes; it must include even the means of protecting metal from rust.

Soldiers who have cleaned grease from stored rifles will recognize the VCI bag as one of the great military advances of our time. It is made of treated paper, cloth and aluminum foil. Rifles and other equipment heat-sealed into it for storage will be clean, rustless and ready for *immediate* use years later.

Perhaps the most astonishing thing to be included in modern weapons is personal *armor!* As "tin hats," helmets came back into use in World War I. The idea of protective body armor has long attracted moderns who never wore an iron suit. Frightened Civil War conscripts bought

"bullet-proof" vests which weren't an adequate protection from a stiff breeze. Considerably better are the two types of really light, strong body armor now in use. One kind is thickly quilted nylon, like an ancient archer's jacket; the other is a "brigandine" with many plastic disks in little pockets. Neither is proof against a direct hit, but they add to the sense of security, they will stop flying fragments and spent bullets and they are credited with saving lives in Korea.

Most recently Army Ordnance has demonstrated the largest mobile field artillery gun ever known. This has been called the atomic gun because it is capable of firing a shell with an atomic charge and has successfully done so, but it can also shoot ordinary shells and is actually a long-range, general-purpose gun which has been made as mobile as possible. It will shoot twenty miles. On the road it travels suspended between two huge motor trucks, each with its own driver. The two drivers are able to talk to each other by telephone, so they'll both take the same road. The trucks are equipped not only to haul the big gun but also to lower and lift it. With their help it can be set up ready to fire twenty minutes after it arrives at a new emplacement.

The U. S. unveiled the 2.36-inch Rocket Launcher in Africa during World War II. The G.I.'s took one look at its striking resemblance to Bob Burns's gas-pipe horn and instantly christened it "bazooka." Except in official paperwork it will never be called anything else. A bazooka can be

3.5-INCH SUPER-BAZOOKA

57 MM. RECOILLESS RIFLE

handled by two men. One holds it on his shoulder, aims it and fires it; the other attends to the loading. It weighs only about ten pounds but it can knock out a light tank several hundred yards away.

The business end of a bazooka is nothing more than the old Fort McHenry rocket brought up to date and packing a twentieth-century punch. The gas-pipe part is a launcher which can be quite accurately aimed and which is equipped with a shoulder stock. In the hand grip and operated by the trigger is nested an electrical primer to spark off the rocket. The tube is wide open at both ends. There's no recoil whatever, nor is the force of impact on the target of any importance. All it has to do is get there.

The bazooka rocket has what is known as a "shaped charge." Without attempting a technical explanation, this means it successfully concentrates most of the force of the explosion on a very small area. The effect is to punch a hole of approximately one inch diameter in armor plate, which

may be as deep as ten inches; and in hardened steel that's deep! The metal blown out of the hole is sprayed with deadly force on the far side of the armor—that is on the *inside* of a tank, where it sets fire to stored shells and fuel.

The original bazooka has been joined in the field by a three-and-a-half-inch big brother, known as the super-bazooka. It too is a close-up antitank weapon but it can reach farther than the 2.36-inch and it's more accurate. The rocket it fires is heavier and longer and, of course, more powerful. Ordnance says, "The shaped charge in this rocket enables the round to penetrate any armored vehicle now known." Its first seven shots in Korea knocked out seven Russian tanks!

Not so much noise has been made about recoilless rifles as about bazookas, yet in some ways they are even more remarkable. Recoilless rifles are classed as "small arms" because they are individual infantry weapons, yet they fire the same size shells as field guns of 57 mm., 75 mm., and 105 mm.!

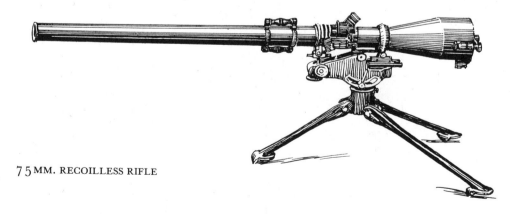

75 MM. RECOILLESS RIFLE

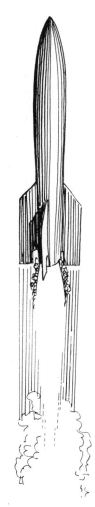

V-2 ROCKET BOMB

The 57 mm. rifle weighs forty pounds and can be fired from a man's shoulder or from its own mount; the 75 mm. fires from a machine-gun tripod; and the 105 mm. (that's more than four inches) can be and usually is mounted on a jeep. These guns have absolutely *no* kick. The cases of their shells are closely perforated to allow the gases to escape to the rear instead of kicking against the breech of the gun. There is space around the cartridge case in the chamber and the breech block is partly open. The force of the explosion goes both ways: it pushes the projectile forward, and it pushes the air rearward—and it pushes hard. Immediately behind the gun is not the best place to stand.

These are single-shot guns and, like the bazooka, are loaded by one man and fired by another, who can't shoot until the loader has released a safety. The recoilless rifles have slightly less range than some artillery of similar sizes but the loss is more than made up by lightness and handiness. Never before has it been possible to put such hitting power into the hands of foot soldiers.

Large rockets as very long-range weapons first appeared late in World War II. Strictly speaking, the first one, the German "Buzz Bomb," was a jet-propelled missile taking in its oxygen from the surrounding air, not a rocket which carries its own oxygen. The Buzz Bomb was about twenty-five-feet long and it had wings with a sixteen-foot spread. Though it carried half a ton of explosive and with its shrieking approach was able to terrify the strongest, it was a military flop. Its launching platforms were easy to find and bomb. The

missiles themselves were easy to track on radar screens and, since they flew only three hundred miles an hour, the English were able to catch and explode eight out of ten of them before they had crossed the Channel. Many were downed by anti-aircraft fire using shells fused with "proximity" fuses which detonated without actual contact.

The V-2 which followed was another thing entirely. It had no wings. It was a true rocket traveling on alcohol and liquid oxygen and reaching speeds above three thousand miles an hour. Since that is noticeably faster than sound, the V-2 did no shrieking. It arrived and exploded its 2,200 pounds of TNT before anybody knew it was coming. Even if radar spotted one, there were no planes which could catch it and no guns which could shoot it down.

After the war, Army Ordnance brought all of the captured German V-2's (about a hundred) and many of the German rocket scientists to the United States, and set up a long research program on guided missiles which has included test-firing many V-2's and other big rockets. One of these last, a "WAC Corporal," went higher than man has ever sent anything, 250 miles, by riding a V-2 to the top of *its* flight and taking off from there under its own power—the "bumper" method they call it.

From these experiments have come several guided missiles designed for special uses. These are reported to give startling performances about which the Army has nothing to say. Each of the three services sees a different need for the guided missile and each has experimented along the

lines of its need. The Air Force is probably most interested in a projectile which can be controlled from a "mother plane" to attack enemy aircraft —"air-to-air" they call it. The Navy has announced its pilotless missile "Regulus," which can be launched from a ship and has actually been launched from a submarine. None of these projectile planes is intended to return from a mission; all carry explosive war-heads by which they are destroyed along with their targets.

Apparently none of the guided missiles has thus far (1953) been used in actual combat, though the Navy did put on a show in Korea using an obsolete plane controlled by radio. In addition to bazookas, all three services make use of rockets. Some, like the "Mighty Mouse" of the Air Force, are very effective but neither it nor any of the others seems to be capable of having its flight direction changed after it has been fired.

The most dramatic hand weapon of World War II and one of the most effective for its purpose was the flame-thrower. It is to be hoped that it is less frightfully inhuman than it seems. It is said to kill its quarry almost instantly. There's nothing to it but mixed oil and gasoline under pressure—and a hose with a nozzle. The stream of flame can be adjusted like the stream of a garden hose. The flame will enter even the smallest porthole in a tank or a concrete "pillbox" and will instantly consume all the oxygen inside. Usually flame-throwers are one-

man weapons, but sometimes they are mounted on tanks.

World War I military aviators made up air bombing as they went along, and the bombs they used were invented from day to day. Between the wars the shapes of bombs were improved and so was their destructive efficiency. Most of the bombs dropped in World War II were either demolition bombs or incendiaries. The first were thin-shelled and contained high explosives of some kind. Their object was to destroy buildings and supply centers by blowing them apart. They were usually provided with contact fuses which were "armed" by movement in the air. As the bomb fell, its slipstream spun a little pinwheel completely off its threaded stud. This removed a restraint from the firing pin *after* the bomb was safely away from the plane. Demolition bombs come in sizes from a hundred pounds up to ten-ton "blockbusters." Army Ordnance had made, by the end of World War II, a number of bodies of a monster 42,000-pound bomb, but no plane then available could get it off the ground.

Incendiary bombs contain some substance which will burn with intense heat and be more than ordinarily difficult to extinguish. The most recent and one of the most effective incendiaries is jellied gasoline, known to the trade as *napalm*. It burns fiercely and tends to cling where it lands. Its use is mostly against ammunition dumps and truck convoys.

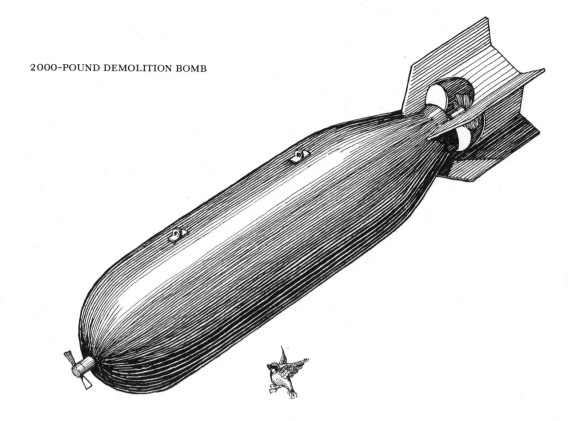

Fragmentation bombs which break up like grenades have been used against troops; so have bombs filled with poison gas, but not since 1918. Bombs can be loaded with disease-carrying bacteria and actual preparatory work has been done along these lines but no such bomb has ever been dropped, and it's a safe bet that none ever will be. Pestilence spreads too easily and too far. Prisoners would bring it back to the troops who started it; an advance into enemy country where invaders had sown plague could wipe out the sowers.

High-altitude bombing is of two types, pattern and precision. The first is like shooting with a shotgun, the second like using a rifle. In pattern bombing, the area to be harassed is sprinkled with many bombs, so that every part of it is likely to suffer some damage. The ultimate in pattern bombing is "saturation bombing" in which the target is so liberally pelted that no part of it is unhurt.

Precision bombing from high altitudes is an American specialty. The fine Norden bomb sight used in World War II made nothing of the intricate calculations needed to allow for speed, altitude and wind drift, and placed bombs on remarkably small spots, from levels high enough to be nearly above the reach of antiaircraft fire.

Not all bombing is done from such heights. Dive bombing and skip bombing are likely to be quite accurate. Both are done by releasing a bomb at low altitude and are not without their special risks.

To end a modern book about weapons without mentioning the atomic bomb and the more powerful hydrogen bomb would be absurd. To try to explain them would be even more ridiculous. Statistics mean little. The bombs are tremendous and uncomfortable facts, and that's about all that can be said of them with certainty. A defense we can't now imagine may neutralize them in time. We can only hope that man will have sense enough to ban them before it is too late.